Preserving Government Information:
Past, Present, and Future

by James A. Jacobs and James R. Jacobs

FreeGovInfo Press

Preserving Government Information

FreeGovInfo Press
San Diego, San Francisco
Copyright 2025

Suggested citation: Jacobs, J.A. and Jacobs, J.R. *Preserving Government Information: Past, Present, and Future*. San Diego: FreeGovInfo Press, 2025. https:freegovinfo.info/PGI/

Names:
Jacobs, James A., (author) https://orcid.org/0009-0009-9231-5512
Jacobs, James R., (author) https://orcid.org/0000-0002-6375-5940
Title: Preserving Government Information: Past, Present, and Future
ISBN: 979-8-218-66931-7 (paper)
Library of Congress Control Number: 2025908228

This book is available as a free PDF and ePub at
https:freegovinfo.info/PGI/

Dedication

To Anita Schiller
friend, colleague, comrade, mentor

Acknowledgements

Government information librarianship is inherently iterative and collaborative, standing on the shoulders of giants as it were. And the same goes for this work, as we have garnered support and inspiration from such pillars of the library community as Bernadine Abbott Hoduski, LOCKSS founders David Rosenthal and Vicky Reich, and Clifford Lynch, the director of the Coalition for Networked Information (CNI). This work would also not have become reality without the tireless efforts and support of too many past and present librarians to name but who have all been active in the American Library Association's Government Documents Round Table (GODORT) and the Federal Depository Library community. We owe a great debt to the many librarians and archivists who patiently answered our questions, helped us track down documents, and generously provided us with the data we analyzed. There were many behind the scenes, but we must thank those we corresponded with directly. These include Chase Dooley, Abigail Grotke, and Grace Thomas at the Library of Congress, Marie Waltz at the Center for Research Libraries, Meg Phillips, Rebeccah Baker, and Theodore Hull at the National Archives and Records Administration, Dr. Sawood Alam at the Internet Archive, Kelly Smith at University of California San Diego, and Ashley Dahlen at the Government Publishing Office. Our gratitude also goes to Shari Laster, who patiently read and gave substantive feedback on an early draft of this work. We were also inspired by the many contributors to the Preservation of Electronic Government Information (PEGI) Project, including Deborah Yun Caldwell, Marie Concannon, Dr. Martin Halbert, Lynda Kellam, Scott Matheson, Robbie Sittel, and Katherine Skinner. Many thanks go to Jessica Meigs, our valiant proofreader, and to Deborah Yun Caldwell, who designed our cover. And, of course we could not have finished this book without the patient support of our braintrust, Shinjoung Yeo and Mary Lou Locke.

Epigraph

[T]he problem of preserving digital information for the future is not only, or even primarily, a problem of fine tuning a narrow set of technical variables. It is not a clearly defined problem like preserving the embrittled books that are self-destructing from the acid in the paper on which they were printed. Rather, it is a grander problem of organizing ourselves over time and as a society to maneuver effectively in a digital landscape. It is a problem of building—almost from scratch—the various systematic supports, or deep infrastructure, that will enable us to tame anxieties and move our cultural records naturally and confidently into the future.

   — Donald Waters and John Garrett (1996)

Democracy's Library is about taking information straight from governments and making these materials permanently available to anybody. That is new. Or new again. In some sense, it's old.

   — Brewster Kahle (2022) (Bustillos, 2022)

Digital libraries are constructed—collected and organized—by and for a community of users, and their functional capabilities support the information needs and uses of that community

   — Christine Borgman (Borgman, 2003)

Not least of the effects of industrialism is that we become mechanized in mind, and consequently attempt to provide solutions in terms of *engineering*, for problems which are essentially problems of *life*.

   —T.S. Eliot (Eliot, 1945)

Table of Contents

Introduction

The digital information that the US federal government publishes is not being adequately preserved. Information is being permanently lost every day. This is an odd situation because there is a virtually universal consensus that government information should be preserved, and the government itself has the responsibility for ensuring that preservation.

This problem exists because almost all government information is now published online and the existing preservation infrastructure was designed for paper-based information, not for digital information. The laws, policies, and procedures that are in place, which mostly worked to preserve paper publications, are proving to be demonstrably inadequate to preserve digital information. The digital-age updates to those paper-era policies have exacerbated existing preservation gaps and created huge new gaps. And even the projects that do exist to preserve digital government information suffer from inefficiency, inadequate funding, minimal access, and few long-term guarantees. The existing preservation infrastructure is so inadequate that even the extent of information loss is difficult to quantify. There is no easy way to accurately and comprehensively identify either what the government has published or what has been preserved.

What is needed is a new infrastructure, an infrastructure designed specifically for preserving digital government information. That is a tall order because "infrastructures" are so much more than computers and networks, and because "preservation" is so much more than just storing files. And, as we will show, infrastructures are not built or legislated; they emerge from practices, standards, and social norms. They emerge when many actors use common understandings and shared goals and well-established strategies.

The goal of this book is to begin defining understandings and goals and suggesting strategies. We provide a vision of a new infrastructure, one designed for preserving digital government information. Implementing this vision will require a shared vocabulary of preservation, analysis of existing data, a theoretical framework based on existing experience, and a flexible approach that recognizes the inherent differences between printed information and digital information. We take on this task in the four parts of this book.

Part One: Context. We begin by defining the landscape of government information and information preservation. We explicitly examine why government information needs to be preserved, because one cannot know what to preserve without knowing why it is worth preserving (Chapter 1).

We then define exactly how we use the terms "information," "government information," and "preservation" in this book (Chapters 2-4). This provides a precise vocabulary for examining practices and assumptions in the paper era and the digital age. (We introduce technical terminology for digital information and digital preservation in Parts Two and Three.)

Part Two: Where Are We and How Did We Get Here? With that foundation in place, we describe the existing laws and policies that govern the preservation of federal government information (Chapter 5). This provides the legal context for understanding the very complex and different ways information is produced by and preserved for each of the three branches of government.

Next, we use data from the Internet Archive's holdings of the End of Term Archive (EOT) 2020 web harvest to analyze the quantity and nature of the government's digital publishing (Chapter 6). In the absence of a complete and precise measurement of the government's digital publishing, this provides a preliminary baseline for use in evaluating the state of preservation.

In Chapters 7-12, we analyze the preservation policies and activities of the Government Publishing Office (GPO), the Library of Congress (LC), and the National Archives and Records Administration (NARA). This includes explanations of what they do and do not preserve and the reasons behind their decisions. Using data from GPO, LC and the Internet Archive (IA) we analyze different patterns of preservation for each branch of government and the reasons for differences between branches.

Using those analyses of laws, policies, and data, we describe six significant gaps in preservation (Chapter 13).

Part Three: Preservation Infrastructures. These chapters provide the theoretical foundation for a Digital Preservation Infrastructure (which we describe in more detail in Part Four). We begin by defining the parts of an infrastructure (Chapter 14). We then describe the essential functions of preservation using the terminology of the OAIS preservation standard, the *Reference Model for an Open Archival Information System* (Chapter 15). In Chapter 16, we describe the inherent characteristics of digital information that are different from printed information and the challenges and opportunities these present to preservationists. We conclude this theoretical part of the book by contrasting old, paper-era assumptions about preservation with new assumptions for the digital age (Chapter 17).

Part Four: A Digital Preservation Infrastructure. We begin this part with a description of the barriers to and opportunities for a new infrastructure (Chapter 18). Using those and drawing on the history, data, and ideas presented in Parts One through Three, we outline the elements needed for a Digital Preservation Infrastructure. We conclude by offering an open framework for preserving government information (Chapter 20) and specific strategies for implementing that framework (Chapter 21). We elaborate on how those strategies could be implemented and actions that preservationists can take now (Chapters 22-23). Chapter 24 concludes the book with a look at the stakes of failing to act and the incentives to act.

This book is not technical, though it does address technical issues. We recognize that a Digital Preservation Infrastructure will require many diverse skills and will have to enable collaboration among different communities, different kinds of institutions, and people with different skills and responsibilities. With this in mind, the book is written for a broad community of people with an interest in ensuring that the information produced by governments remains available and usable long into the future. This broad community includes LIS students, front-line librarians and archivists, and managers of libraries and archives. It addresses issues relevant to practicing specialists in areas such as metadata, computer programming, web harvesting, database administration, and web development. It speaks to government workers who produce and publish government information and policymakers who design laws and regulations that affect the production, dissemination, and preservation of government information.

The shift from paper-publishing to digital-publishing has been dazzlingly swift when compared to other technological shifts. This means that some active professionals grew up in the paper era and others in the digital age. We therefore take every opportunity to compare the practices and assumptions of paper era with those of the digital age. We believe that development of a successful Digital Preservation Infrastructure can benefit from an understanding of the similarities and differences of these two environments. We believe that the future of digital preservation can best be built with an understanding of both the paper-era past and the present state of the art. We also base our vision on well-understood principles that have been documented by researchers and preservationists over the last 30 years.

Although the focus of this book is on US federal government information, we believe that the underlying issues are relevant to other governments and levels of governments. We focus on US federal government information because it is an enormous body of information that desperately needs attention and because it has a long and ongoing history of preservation activities that can provide clear examples of issues, successes, and failures that should bring the abstract concepts we describe into clear focus.

PART ONE:
Context

Preservation was simpler in the paper era than it is in the digital age. Today it is necessary to have a clear and shared understanding of what it is that we preserve when we preserve information and what it means to say something is successfully preserved.

Preserving government information is also more complex than preserving other kinds of information because government information includes everything from unique archival records, to widely distributed publications like the *Federal Register* and the *Congressional Record*, to databases and raw data like Census microdata.

In Part One, we address the basic context for the preservation of digital government information.

CHAPTER 1
Why Preserve Government Information?

It is almost a truism to say that government information should be preserved. Indeed, the need for preservation is more often simply accepted as a given rather than explicitly justified. Citizens, journalists, and academics know intuitively that free, long-term access to government information is necessary so that citizens can hold their government accountable for its actions. Even private sector businesses know that the information produced by governments is an incredibly rich resource for understanding the country, its resources, and its people.

This kind of implied acceptance of the need to preserve government information was sufficient to justify preservation in the print era because preservation was a passive byproduct of existing policies that focused on access to books. The infrastructure for providing access to government information as it was created had the additional benefit of preserving that information over time in hundreds of public, university, college, law, and government libraries throughout the country.

The technologies of the digital age (digital publishing, the internet, and the world wide web) are the new infrastructure for information access, but not for preservation. While the old, paper-based access infrastructure had preservation as a byproduct, the digital access infrastructure does not. Digital preservation requires a much more active and intentional approach than preservation of paper. Digital preservation requires "the continuous, active management of digital objects," as Peter Hirtle, who has served as Director of the Cornell Institute for Digital Collections and as the Associate Editor of *D-Lib Magazine*, has said (Hirtle, 2008). Doing this will require a new digital-preservation infrastructure.

Creating a digital preservation infrastructure will take more than providing short-term access. It will require long-range planning and clear, long-term goals. Setting those goals will require an understanding not just of what we *can* do but *why* we do it so we can identify *what* to do. Almost 50 years ago the well-known systems scientist Hasan Ozbekhan wrote that the first step toward developing practical goals is not deciding what we can do but having an explicit understanding of what we ought to do (Ozbekhan, 1968). In the print era, we preserved government information easily enough and did not have to examine very closely why we ought to do so. Providing access resulted in preserving information. In the digital age, without an explicit commitment to preservation, access can evaporate overnight. To make that commitment, we need to begin with a clear understanding of why we must preserve government information.

As a starting point to finding that understanding, we can examine the justifications used for preserving government information in the print era. Many preservationists are still using these justifications in the digital age.

Why Did We Preserve Government Information?

In the print era, there were three justifications that were often cited for preserving government information: the importance of an informed citizenry, the value of government information, and preservation as a "principle." All these reasons are good as far as they go. None is incorrect, but all are inadequate in the digital age.

Informed Citizenry

There is a tradition when writing about government information to begin with a reference to Franklin or Madison or Jefferson to justify its importance. For example, the Congressional Office of Technology Assessment (Office of Technology Assessment, 1988) and the National Commission On Libraries And Information Science (US National Commission On Libraries And Information Science, 2000) have asserted the importance of government information by citing Jefferson as saying, "If we are to guard against ignorance and remain free, it is the responsibility of every American to be informed." The Government Publishing Office (GPO) has a bust of Benjamin Franklin in its lobby and cites him as an early "Publick Printer." And GPO—which changed its name from the Government Printing Office in 2014—includes the following famous James Madison quotation every year in its annual *Budget Justification*.

> A popular Government, without popular information, or the
> means of acquiring it, is but a Prologue to a Farce or a
> tragedy; or perhaps both. Knowledge will forever govern
> ignorance. And a people who mean to be their own
> Governors, must arm themselves with the power which
> knowledge gives. (Madison, 1910)

Such general statements that assert the importance of government information and that link access to information with an informed citizenry are laudable sentiments and they make good slogans, but they are too vague to guide policies or actions. (It is notable that some of these truisms are even misleading or unverifiable. Though Franklin did do official printing for the colonies of Pennsylvania and Delaware, he did so as a commercial printer and he died more than 70 years before GPO was created [MacGilvray, 2006]; the famous Jefferson quote about "responsibility of every American to be informed" cannot be found in any of his writings [Thomas Jefferson Foundation, {n.d.}].)

Nevertheless, the implications of these assertions are valid. Citizens cannot be informed of the actions of their government without access to a record of those actions. Preservation of that record is an essential step in making that possible. These assertions imply, but often leave unstated, *why* government information is important enough to preserve. Government information is important to an informed citizenry because *it provides an official record of the activities of government*. That means that any preservation plan will need to ensure the preservation of the official record of activities of government.

Value

Another common justification is that government information is valuable. Again, the implication is clear: one does not want to lose or discard something valuable. "Value" is even written into the law that defines what the National Archives and Records Administration will preserve. One of the criteria for defining Records in the *Federal Records Act* is their "informational value" (44 USC 3301). But "value" as a criterion for preservation is subjective, as we explore in more detail in Appendix C. Even T.R. Schellenberg, whose *Appraisal of Modern Public Records* has guided NARA since 1956, called testing a record for its informational value an "imponderable." Schellenberg asked, "[W]ho can say definitely if a given body of records is important, and for what purpose, and to whom?" (Schellenberg, 1956).

But "preservation" is not the only way of dealing with valuable things. Scarcity, for example, may increase value but lead to limitations on access. Measures of value can drive decision making, which can lead to "value" being used to challenge preservation rather than guarantee it. Preservationists may be challenged to demonstrate how valuable any particular government publication is and compare the cost of preservation to that measurement of value.

Focusing on "value" as monetary value can also shift the focus to costs of production and distribution and preservation. It can invite measures of "return on investment" and lead to commercialization, imposing fees or use-restrictions, privatization, or some combination of all three. Commercialization can put preservation at the mercy of market value. Such actions are not likely to promote or facilitate long-term free access to the complete records of government. They are more likely to lead to preserving only that information that can return a profit (Schiller, 1991). This would be unlikely to create an "informed citizenry."

To be clear, we are not saying that everyone who says that government information is valuable is implying that the information should be treated as a commodity or commercialized. But these

9

arguments have been made repeatedly over the years by, for example, the Office of Management and Budget (OMB, 1985), the information industry (Stiglitz, Joseph E., Orszag, Peter R. Orszag, Jonathan M., Computer & Communications Industry Association, 2000), the National Commission On Libraries And Information Science (US National Commission On Libraries And Information Science, 2000), the House Committee on Appropriations (House Committee on Appropriations, 2011), and the National Academy Of Public Administration (National Academy Of Public Administration, 2013) among others. In 2017, a draft bill to amend Title 44 of the *US Code* would have allowed Congress to privatize congressional publications (GPO, 2018b). As recently as 2020, the Heritage Foundation criticized a policy of the Office of Science and Technology Policy (OSTP) of making federally funded research results freely available (Mossoff, 2020). These repeated arguments cannot be ignored.

We suggest that, rather than just saying we should preserve information because it has value, we should specify the value. To put it simply, *government information has inherent value.*

This inherent value comes from two characteristics of government information. First, the information governments produce is valuable because it comprises a record of the actions of government. That record is essential in a democracy because it is information needed by citizens to participate in the democracy. It is what the "informed citizenry" needs to become informed. Such information has to be preserved to provide context for new information as it is created.

Second, governments collect, compile, aggregate, and create information that is *irreplaceable*. This is certainly true of obvious creations such as laws, regulations, and judicial decisions. But it is also true of administrative records (e.g., SEC filings, birth and death records), surveys (e.g., censuses), and economic indicators and data. If not preserved, that information will be lost because it cannot be recreated later.

There is one other point we need to mention here briefly. This is that the inherent value of government information is only *potential* value. Its value is only actuated when it is usable and used. Information that cannot be found or cannot be used is like a tractor locked in a barn; it may have potential value, but its real value is only realized when it is taken out of the barn and used to plow fields. When its use-value is locked behind paywalls because the information is "valuable," its use is restricted to those who can afford to pay for it. This use-value is therefore an essential component of preservation; we will return to this idea in Chapter 4 when we define preservation.

In summary, the assertion that government information should be preserved because it has value is accurate, but as a practical matter in preservation planning, this simple assertion is inadequate. Assertions of the value of government information must explicitly include an understanding of the inherency of that value and the irreplaceable nature of the information in order to justify and guide its preservation.

Principles

Since at least the 1980s, various organizations have expressed their "principles of government information," including responsibilities for preservation. These "principles" are somewhat like manifestos or resolutions in that they express the values of the organization that created them, but they are generally not specific enough to be used as goals for preservation. Such principles are also usually expressed by those without the authority to create policy and are used to suggest that others should assume that responsibility. Each of these sets of principles says that preservation is needed and at least hints at the need to preserve digital information. Although these principles do not say *why* government information should be preserved, they do often suggest *who* should preserve it.

The Office of Management and Budget (OMB) included principles in the first version of its *Circular A-130* (OMB, 1985). These were noteworthy because of their explicit focus on government information as a "commodity with economic value" and OMB's insistence on reliance on the private sector for dissemination. OMB tied the "value" of preserving government information to "the legal and financial rights of the government or its citizens." (Interestingly, the draft also provided a pretty good statement about the inherent value of government information, saying it "provides an official record of Federal agency activities for agency management, public accountability, and historical purposes.") OMB had the authority to enforce its principles, and this first draft led to controversy and opposition (Hernon, 1986).

The Association of Research Libraries (ARL) issued principles in 1988 suggesting the importance of both "no-fee access" and making government information available through "entrepreneurship" at "low cost." Its only mention of preservation was this: "Federal policy should support the integrity and preservation of government electronic databases" (Gapen, 1989).

In July of 1990, the National Commission on Libraries and Information Science (NCLIS) issued eight "Principles of Public Information," including, "The Federal Government should guarantee the integrity and preservation of public information, regardless of its format" (US National Commission On Libraries And Information Science, 1990).

The American Library Association's Government Documents Roundtable (GODORT) issued a draft of 11 principles in 1990 (GODORT Committee on Legislation, 1990). This short statement (less than 100 words) closely paralleled NCLIS's eight principles but added explicit references to "electronic" information and keeping government information free of copyright. The draft also included a principle on preservation: "Government has an obligation to archive and preserve public information, regardless of format." A year later, GODORT added 1,000 words of "annotations" to the principles (GODORT, 1991). By 2004, the principles along with their annotations had been officially approved by a number of ALA groups, including the Committee on Legislation, the Public Library Association (PLA), and subcommittees of

Library and Information Technology Association (LITA) and the Association of College and Research Libraries (ACRL) (GODORT Committee on Legislation, 2004). ALA has kept these principles on its government information web pages since at least 2008 (American Library Association, 2025).

In 1996, GPO issued its own five "principles for federal government information" based on the 1990 NCLIS principles. GPO dropped explicit references to "electronic information" although the principles were part of a GPO's *Study to Identify Measures Necessary for a Successful Transition to a More Electronic Federal Depository Library Program* (p.4-7) (GPO, 1996b). GPO did not include mention of the utility of information, the Federal Depository Library Program (FDLP), government funding, privacy of users, or the need for a catalog—all factors that GODORT, NCLIS, and even OMB had included. (The omission of these aspects may say more about the vagaries of "principles" than it does about GPO's values.) GPO's preservation principle was simply that "The Government Has an Obligation to Preserve Its Information." Later that year, ALA's Government Information Subcommittee of the Committee on Legislation (COL) and GODORT's legislation committee passed a resolution endorsing GPO's principles (GODORT Committee on Legislation, 1996). Unfortunately, saying that government has an obligation has not made it happen. We will see in subsequent chapters how such an obligation plays out in the face of weak and conflicting laws, inadequate budgets, and policies based on convenience rather than on goal-driven outcomes.

In 1998, GPO issued a separate set of "broad principles and core values" as part of its planning document, *Managing the FDLP Electronic Collection* (GPO Library Programs Service, 1998). Unlike GODORT and NCLIS and ARL and even OMB, GPO had the power to implement its principles. Its most notable contribution to the genre of principles was asserting that GPO was given responsibility for "providing permanent public access" to electronic government information by the *Government Printing Office Electronic Information Access Enhancement Act of 1993*.

(This assertion was not strictly true, but this interpretation has driven GPO policies, as we shall see in Chapter 5.)

In 2016, GPO missed the opportunity to revise or update its 20-year-old policies and, instead, reaffirmed its commitment to its 1996 principles, explicitly expressing the assumption that "[t]he principles of Government information adopted by GPO in 1996 remain valid for the digital age" without mentioning its 1998 principles (GPO, 2016c).

All these principles have two things in common with regard to preservation of government information. They do not address the question of *why* we preserve government information, but they do address the question *who* should preserve government information: the government. Using the terminology of preservation planning (see Chapter 19), these principles set a goal of "preserving government information" and a tactic of "the government" doing the preservation. But "preserving government information" is a vague goal and inadequate if there is no explanation for why it should be preserved. Choosing a tactic (who should preserve government information) before developing a strategy (what should be preserved and how) is simply bad planning. Surely the government has a role to play in preservation of its information, but its role should be based on a goal-driven preservation strategy, not on a simple assertion of a "principle."

Why Government Information Should Be Preserved

To repeat, those three traditional reasons for preserving government information are not wrong; indeed, they are correct as far as they go. But they do not go far enough. They fail as justifications for preservation because they are too vague to drive any goals or evaluate any policies. This brief bit of history of those three reasons does, however, suggest three more explicit justifications for preserving government information that can drive goals and be used to evaluate the success of policies.

- Government information has inherent value as the official record of the actions of a government and so must be preserved as an essential prerequisite to informed public participation in a democracy.

- The information that the government collects, compiles, aggregates, and creates in the course of its normal statutory and regulatory functions must be preserved because it comprises irreplaceable knowledge about and for the republic.
- The inherent value of government information is realized through its accessibility and usability by the general public.

CHAPTER 2
What is Information?

> Before Shannon, information was a telegram, a photograph, a paragraph, a song. After Shannon, information was entirely abstracted into bits. (Soni & Goodman, 2017)

What exactly is it that we preserve when we preserve government information? Although this might sound like a question with obvious answers (e.g., a book, a PDF file, a web page, etc.), the definition of "information" in the digital age is more complex and subtle than it was in the print era and that subtlety carries critical implications for digital preservation. These implications (which we will examine in Part Three and in Appendices C and D) include effects on what we select to preserve and how we preserve and deliver information. Claude Shannon is often called "the father of the information age" and the quotation above, from the biography *Man in a Hurry*, encapsulates the issues we address in this chapter: what do we preserve when information has become abstracted into bits?

What Information Is

The word "information," like so many words, has many meanings. Everyone who has studied or worked the field of library- or archival- or information- or computer-science, or who has been actively involved in the preservation of government information, probably has at least an intuitive working definition of "information." The problem is that each field uses a different definition. As Donald Case said in his book, *Looking for Information*, "The distinctions and disagreements among reviewers of definitions are too many to resolve; in short, there is as yet no single, widely accepted definition for the concept of information" (Case, 2002).

Or, as Jonathan Furner, UCLA professor of Information Studies, has said, there are as many definitions of information as there are writers on the topic (Furner, 2015).

Our task here is not to resolve all those differences but to define "information" specifically for use by those preserving government information. This is important and necessary in order to enable people with different training and experience to communicate clearly with each other and work together without misunderstandings.

Our starting point is that, simply as a practical matter, you cannot preserve something unless you have a thing to preserve. When you read that "the amount of information on Earth is doubling every two years" (Zegart, 2022) and similar claims, the volume of "information" is not words or images or ideas or knowledge; it is words recorded, images saved, ideas written down. It is material stuff—whether the material is paper or disk, whether the stuff is ink or bits.

The Office of Management and Budget (OMB), in its guidelines for *Circular A-130, Managing Information as a Strategic Resource*, defines information as "any communication or representation of knowledge such as facts, data, or opinions in any medium or form, including textual, numerical, graphic, cartographic, narrative, electronic, or audiovisual forms." OMB gives us a good theoretical starting point by describing information not as "knowledge" but as a *representation* of knowledge (OMB, 2016).

Michael Buckland differentiated "knowledge" from its representation. He defines the abstract concept of knowledge as "information-as-knowledge" and the representation of knowledge as "information-as-thing." Buckland, who served as dean of the School of Library and Information Studies at Berkeley and assistant vice president for library plans and policies for the University of California, described *information-as-knowledge* as facts or opinions, and *information-as-thing* as "the objects, such as data and documents" used to transmit and store information-as-knowledge. Buckland explained that to communicate or transmit information-as-knowledge, it has to be "expressed, described, or represented in some physical way, as a signal, text, or communication." This is almost identical to OMB's definition. Buckland described these representations of knowledge as being "tangible." He was not using the word tangible to refer to physical media, but to the instantiation of knowledge into a "thing" that could take many forms: "sign, signal, data, text, film, etc." This is very similar to OMB's qualification that those representations could be "in any medium or form, including textual, numerical, graphic, cartographic, narrative, electronic, or audiovisual forms." Buckland's definition tells us that these are all objects (whether physical "documents" or "signals" or "data") that are tangible representations of information-as-knowledge. In contrast, information-as-knowledge is always intangible. Significantly, Buckland said of intangible knowledge that "one cannot touch it or measure it in any direct way," but information systems (in other words, the computers that we use to manage digital preservation) can deal directly *only* with "information-as-thing" (Buckland, 1991).

Buckland has also described the evolution of the term "document" in his often-cited paper, "What Is a 'Document?'." Buckland traced the history of the word "document" from its earlier use as a reference to texts and books, to its use to refer to any describable information object. He quotes theorists from 1930s and 1950s. Walter Schürmeyer said a document was "any material basis for extending our knowledge which is available for study or comparison," and Suzanne Dupuy-Briet called it "any physical or symbolic sign, preserved or recorded, intended to represent, to reconstruct, or to demonstrate a physical or conceptual phenomenon." Buckland summarizes this, saying that a consensus was emerging that a "document" was whatever *functioned* as a document regardless of its physical form (Buckland, 1997).

This distinction between the abstract idea of knowledge and the representation of knowledge in physical or digital objects is not a radical new idea. This is simply a practical matter of information science that has a long history. Paul Otlet described this understanding more than 100 years ago. Otlet was one of the first information scientists at a time when the field was primarily concerned with the management of "documents." He created the Universal Decimal Classification and co-founded the International Institute of Bibliography, which evolved to become the International Federation for Information and Documentation. In 1903, he distinguished between "physical, concrete object[s]" and "the ideal, abstract concept[s]" contained in those objects. He even developed a hierarchy of information with "Knowledge or Understanding" or "everything we know" at the top. This was the abstract, intangible concept of knowledge, which could be represented as a document, or writing, or printed work. At the bottom of his hierarchy were "Books," which he described as "printed works which are published separately" (Otlet, 1990).

Otlet's and Buckland's ideas are not unique. As other reviews of the literature have demonstrated, this simple idea is mainstream in information science. Donald Case devotes an entire chapter of his book to examining definitions of information and a section of that chapter to definitions that include "physicality" as a requirement and how these contrast with the concept of knowledge as intangible. McCreadie and Rice in their review of access to information describe research that singles out the concept of information as a "representation" of knowledge. They note that "representations" have traditionally referred to documents, books, and periodicals but increasingly refer to digital alternatives (McCreadie & Rice, 1999). Marcia Bates, in her lengthy entry on Information in the *Encyclopedia of Library and Information Science*, finds the same concepts of intangible knowledge and tangible representations of knowledge in philosophy (Popper) and a study of the evolution of information transfer (Goonatilake) (Bates, 2018).

This understanding of information as the instantiation into a storable and transmittable tangible form of an intangible "work," of "a distinct intellectual or artistic creation," of "conceptual content," of a "story," of "ideas in a person's head," is what underlies the modern *Functional Requirements for Bibliographic Records* (FRBR) standard and its successor, the *Library Reference Model* (LRM) (IFLA Study Group on the

Functional Requirements for Bibliographic Records, 2009; Tillett, 2004). Indeed, the LRM notes that no "work" can exist without there being at least one "expression" of the work (Riva et al., 2017).

This understanding of information has practical applicability to its preservation and helps define the inherent differences of preserving printed information and digital information. In the print era, the representation of information was the instantiation of words into ink on paper in physical objects: books. Once instantiated into a book, the physical object became the target of preservation. Preserve the book and you preserve the information.

But that model fails in the digital age because the representation of digital information is not a physical object. The representation of digital information is its instantiation into bundles of bits that we call digital objects or (less precisely) "files." (The concept of "the file" is complex and really more of a metaphor than a physical thing [Harper et al., 2011; Lindley et al., 2018; Banerjee & Forero, 2017]. Indeed, students in the 2020s may not even use the term "file" anymore [Chin, 2021]. It is, nevertheless, still a convenient and familiar label for many of the things we preserve.) There is "information" to preserve only when content is encoded into these bundles of bits. Those bundles, those digital objects, are the tangible representations, the "thing" that we preserve. The digital object can be (and, over its lifetime, will be) stored in any of a number of different physical digital media. A file can be stored on different kinds of tape or disk or on the physical memory chips of a computer. Unlike the printed book, the file's storage is not tied to any single storage medium and, so, the information is not tied to any single medium.

This reveals a new and defining characteristic of digital preservation. Since the digital information is no longer tied to a single medium, preservationists no longer manage physical objects; we manage information itself. In the print era, we managed and preserved physical media (books, etc.); in the digital age, the physical digital media objects (tapes and disks, etc.) are simply temporary, disposable and replaceable tools that we use to store the target of preservation: the digital object, the file. The physical digital media come and go, change and evolve, are discarded and replaced, but the information abides.

It is worth mentioning briefly that the digital shift not only affects storage; it also affects the other two functions we typically require for information: transfer and use. In the print era, the book served all three functions. We preserved the physical object. We transferred the same object to users of the information, who used the same physical object to access the information (read the book). Remarkable! But in the digital age different media can be used for each of those functions. We might receive a file on tape, store it on disk, and deliver it to a user over the internet who might use it by printing it onto paper. During preservation, we might migrate a file from old disks to new disks. It is always the digital object, not the physical medium, that is preserved, stored, transferred, and used. The different media are just tools for those functions. This functional split opens remarkable new opportunities for information exchange, use, and reuse, and we will examine the implications of that in Chapters 16 and 17 and in Part Four. The decoupling of information from its containers is one of the defining characteristics of digital age information. It affects every point in the life cycle of information. Successful digital preservation depends on understanding and making use of this feature.

We belabor this point at the risk of stating the obvious. It is, or should be, obvious, but even when well understood in principle, it can be misunderstood in practice. It is a mistake, for example, to confuse the information preserved with the medium used to transfer or store that information. That confusion is the source of the common and unfortunate use of the phrases "intangible information" (i.e., information available online) and "tangible information" (i.e., information available on portable media such as DVDs) as if they were kinds of information. Those are categories of media, not categories of information. This confusion has had a direct effect on digital preservation, which we examine in Appendix B.

21

All this leads us to a working definition of information based on Buckland's definition.

> **"Information"** or *"Information-as-thing"* means the object, such as a book or a digital object, that is stored, transmitted, and used to communicate "information-as-knowledge."

We offer this not as a substitute for other definitions but as the meaning we use in this book when we refer to preserving information. Knowledge and thoughts and ideas and opinions and facts become "Information" that can be preserved when they are instantiated as a tangible information-as-thing, whether that "thing" is a physical object such as a book or a string of bits.

CHAPTER 3
What is Government Information?

Now that we have a working definition of information, we can ask the more specific question: What is government information? In order to preserve it, we have to be able to understand what it is. This will help us consistently identify the universe of information that *should* be preserved. It will help to understand the legally binding terminology that determines who has responsibility for preservation and what information is available for preservation.

The digital age has blurred some of the traditional distinctions between kinds of government information. Where the difference between "records," which went to the National Archives and Records Administration (NARA), and "publications," which went to GPO to be distributed to Federal Depository Libraries (FDLs), was fairly clear and mostly unambiguous in the paper era, the difference between a "record" and "public information" and who will preserve each is not well defined or understood in the digital age.

It is important to have consistent, clear terminology so that everyone from technologists to managers to legislators can discuss preservation productively.

A Hierarchy of Government Information

There are many ways of categorizing the universe of government information, and each has its own utility. We begin here by using categories drawn from official federal government definitions in the *United States Code* (USC) and *OMB Circular A-130* (OMB, 2016). The *USC* provides the legal definitions used for US information policy. *Circular A-130* provides guidance to executive branch agencies in their implementation of official policy. These definitions are useful because they are official and have some weight in determining what is available for preservation—and what is not. They are, therefore, both essential and useful to the practicing preservationist. In some cases, these short definitions are not as precise as one might wish, but they provide a good starting point for developing a more precise, consistent preservation vocabulary.

We have selected six terms that are the most relevant to preservation from the many that are defined by the *USC* and *A-130*. We organize these six terms into a hierarchy of categories—each one consisting of a subset of the category above it. Although neither *USC* nor *A-130* explicitly describe this hierarchy, we believe the wordings of the definitions imply it. The references after each definition are to the title and section of the *USC* and the numbered definitions in the 2016 edition of *Circular A-130*.

> Information
> ... Federal Information
> ...... Records
> .... Public Information
> .... Information Dissemination Product
> .... Government Publication

Information

> "Information" means any communication or representation
> of knowledge such as facts, data, or opinions in any medium
> or form, including textual, numerical, graphic, cartographic,
> narrative, electronic, or audiovisual forms. [OMB 28]

This is the definition we quoted in Chapter 2. This leads to our understanding that, for purposes of preservation, information is not knowledge or facts or opinions but the *representation* of knowledge, facts, and opinions. This is a useful category because it is broad enough to encompass all the different kinds of information that the government collects, creates, uses, and disseminates and the different formats and media used in storing, transmitting, and presenting them in the 21st century. It defines information as either a "communication" (transmission) of or a "representation" (storage) of knowledge, facts, data, or opinions. OMB uses "electronic," which, presumably, includes electronic media such as analog audio and video recordings. The definition explicitly includes information stored or communicated using any "medium or form." Though it neglects to define what it means by these two terms, we can infer that the intent of the definition is to be all inclusive. The absence of any exclusions in the definition reinforces this inference.

Federal Information

> "Federal information" means information created, collected, processed, maintained, disseminated, disclosed, or disposed of by or for the Federal Government, in any medium or form.
> [OMB 22]

Of all the Information in the world, a subset is Federal Information. This is the information that the United States federal government collects and creates. This is an enormous category, and it is important to define subcategories of it. The list in this definition ("created, collected, processed, maintained, disseminated, disclosed, or disposed of") reflects the stages in the life-cycle of information.

Records

> "Records" means all recorded information, regardless of form or characteristics, made or received by a Federal agency under Federal law or in connection with the transaction of public business and preserved or appropriate for preservation by that agency or its legitimate successor as evidence of the organization, functions, policies, decisions, procedures, operations, or other activities of the United States

Government or because of the informational value of data in them [44 USC 3301; OMB 70]

A subset of all Federal Information consists of what archivists call "Records." Indeed, this definition is from the section of *US Code* that deals with records administration and management by the government. The specific components of the definition of "Records" have evolved over time. Essentially, the law says that Federal Records include information that the government receives and creates under law as it transacts official business. Such records are evidence of actions of the government. OMB relies on this same definition, but the *Code* goes further than OMB by also defining "Recorded information" as including "all traditional forms of records, regardless of physical form or characteristics, including information created, manipulated, communicated, or stored in digital or electronic form." It also gives the Archivist of the United States the authority to make binding determinations on what is and is not "recorded information" (44 USC 3301(b)). In practice, executive agencies are given broad leeway in determining which of their Records should be scheduled for preservation.

This is still a very broad category, and to better understand its scope it is useful to briefly differentiate between three different kinds of Records. First, there is information that the government *collects*. Examples include testimony at Congressional hearings, original census forms, judicial hearings, and the crime reports that local police departments submit to the Department of Justice. Second, there are *transactional records* that document the handling and processing of the raw data that the government uses to go about its business, arrive at decisions, or create new information. Examples include meetings of committees of Congress and the debates on the floor of Congress, the data processing that the Census does to create reports, the deliberations of the Supreme Court, and the processing of administrative records into statistical summaries. This category includes "working files, such as preliminary drafts and rough notes" and e-mails (36 C.F.R. 1222.12). Third, there are *outputs* of government actions. These include acts of Congress, published census reports and statistical reports such as *Uniform Crime Reports*, and decisions of the Supreme Court.

Finally, it is worth noting that the definition says that Records can have two different kinds of value: the evidence of actions of government and the "informational value of data in them." These relate to two of the reasons for preserving government information that we define in Chapter 1: the inherent (evidentiary) value and the irreplaceable knowledge collected and created by the government (information value).

Public Information

> "Public information" means any information, regardless of
> form or format, that an agency discloses, disseminates, or
> makes available to the public [44 USC 3502; OMB 68].

A subset of all Records is "Public Information." Not all Federal Information Records are Public Information; some are confidential or secret or private or are otherwise restricted from public access by law. Public Information is those Records that the government makes available to the public. "Records" become "Public Information" when they are made available to the public. In short, some records become Public Information, but most do not.

As we analyze how to preserve government information, our primary focus is on Public Information. But we are also concerned with the preservation of non-public, or unreleased, information because some information that is explicitly withheld from public release will be released (or at least "made available;" see Chapter 17) in the future. For example, the original census schedules submitted by individuals are kept confidential for 72 years but are then, by law, made public (44 USC 2108(b)). Another example of such records are those in the Center for Legislative Archives (NARA Center for Legislative Archives, 2017), most of which are made public after 30 years (House) and 20 years (Senate). Many documents that are confidential or secret at the time of their creation are later declassified and published. For example, the official records of major US foreign policy decisions and significant diplomatic activity such as secret diplomatic cables are eventually published in the series *Foreign Relations of the United States* (US Department of State, Office of the Historian, 2025). Most notably, a large body of Records become Public Information when they are transferred to the National Archives. Many of those become available to the public, even though they are never published or "disseminated" in the traditional sense. Many unreleased Records, while held by agencies or NARA, are subject to requests for release through the *Freedom of Information Act* (FOIA). The Electronic Freedom of Information Act Amendments (EFOIA) defined "Record" to mean "any format, including an electronic format."

In addition to information that is actively withheld for legal reasons, some information that could be released to the public is destroyed because the agency and NARA agree that it has no long-term value. (NARA preserves only 1 to 3 percent of Federal Records [NARA, 2020b].) Other information is withheld passively because the government simply fails to release it. There are many reasons for such decisions (political, budgetary, technical).

We may think of unreleased information that can, should, or will become Public Information as in a kind of limbo awaiting its ultimate release. Such information ultimately belongs to the public but is in the custody and care of the government and will be lost if not preserved by the government before its release.

Information Dissemination Product (IDP)

> "Information dissemination product" means any recorded information, regardless of physical form or characteristics, disseminated by an agency, or contractor thereof, to the public. [OMB. 29]

When information is marked for release to the public, it has to be transformed from its internal storage to a public communication. This is a subtle category, and we will do more to define it in Chapter 15 when we compare it to the definition of a Dissemination Information Package (DIP), a term used by the *Open Archival Information System* (OAIS) preservation standard (ISO, 2012b). It is worth noting the difference between this category and its parent, "Public Information." Public Information includes information that is "available" to the public. IDPs are the subset of all Public Information that has been "disseminated" to the public.

All Federal Information is collected, created, and stored in ways that make it possible for the government to carry out its legally mandated duties. For release to the public, the information has to be converted into an Information Dissemination Product (IDP) appropriate for dissemination for the public. When an agency transforms information from the form in which it is stored in internal systems (e.g., typed reports on paper, databases, email systems, content management systems, word processing documents) to a form for delivering it to the public (e.g., a web site, an XML file, a PDF file, a spreadsheet, a datafile, a posting on social media) it is creating an Information Dissemination Product. In some cases, the IDP is no more than a duplicate copy (a photocopy of a typed page or a copy of a PDF file) of the original. In other cases, the conversion may involve changing the internal storage format (e.g. records in a database) to a different kind of format for the public (e.g. a book, a PDF, a web page, a spreadsheet).

The definition is broad enough to encompass all kinds of "products"—which one might call "packages" or "containers" or "formats"—ranging from simple paper-and-ink books to complex web-based services. (An example of a web-based service is the *Code of Federal Regulations* [ecfr.gov] that incorporates additions, deletions, superseding, and other forms of frequent updating.) OMB explicitly classifies agency websites as IDPs (OMB, 2004).

The earliest use of the term by OMB that we could find was in 1987; OMB incorporated it into *Circular A-130* in 1993 (Office of Technology Assessment, 1988, p. 266; OMB, 1993). This phrase might seem awkward, but it will be useful in discussing the preservation of digital objects such as dynamically produced web pages, entire web sites, databases, and other complex digital objects.

Government Publication

> "Government publication" means information that is published as an individual document at Government expense, or as required by law, in any medium or form [44 USC 1901. OMB 25].

The final sub-category is the one most familiar to FDLP librarians. Historically, the government released and disseminated Public Information to the public by printing "Government Publications" as "individual documents" or, as librarians often referred to them, "government documents." These were mostly printed paper-and-ink publications and, at least since the 1960s, could be on microfiche. These "documents" could take many forms, including books, periodicals, maps, pamphlets, loose-leaf services, and so forth.

The definitions of IDPs and Government Publications are worded differently, but their meanings are virtually identical except for the qualification that a Government Publication must be "an individual document." Neither the *Code* nor the OMB *Circular* define what this means, though. This definition dates back to the *Depository Library Act of 1962*. When that law was written, almost all "government publications" were printed books (or similar "individual documents" like a map or pamphlet) and no further explanation seemed necessary. The significance of this definition, though, is that it defines (however imprecisely) what information "shall be made available to depository libraries" (44 USC 1902). We examine the controversy engendered by this wording below.

It is notable that GPO has interpreted *Circular A-130* as extending the definition of government publication to include IDPs, including "government data sets" (GPO Superintendent Of Documents, 2019a).

The definition we quote here is the OMB definition. It is almost identical to the legal definition (44 USC 1901), except for the last five words, "in any medium or form." As noted above, the Title 44 definition was written in 1962. The OMB version explicitly recognizes the proliferation of media and "forms"—though it neglects to define what it means by those terms. This is a relatively recent change; OMB used the Title 44 definition in the first version of *Circular A-130* (OMB, 1985) and did not add the five words until the 2016 revision of the *Circular*.

Preserving Public Information

In this book, we focus on preserving Public Information and will use that term repeatedly to refer to the universe of content made available to the public by the federal government. This category, as defined in both law and policy, is a better choice than the term "Information Dissemination Product" (IDP) because it encompasses all publicly released government information whether actively "disseminated" or not. The definition of IDP is narrower, encompassing only information that has already been disseminated by being packaged into an information "product." For example, public Records in the National Archives that are not classified or restricted are Public Information even if they are not available on NARA's website, even if they have not been re-packaged for wide distribution, even if using them requires visiting a NARA facility.

The universe of all Public Information thus includes two broad types of information that present different challenges for preservationists. We will refer to these as "published Public Information" and "unpublished Public Information." Published Public Information is information that has been explicitly and intentionally distributed to the public. In the paper era, we called these "publications," but in the digital age, this includes virtually everything posted on publicly accessible government websites. Unpublished Public Information is those non-public Records in agency recordkeeping systems that are transferred to NARA without any restrictions on public access. As noted above, these become Public

Information without being packaged for distribution and without being actively distributed. As we will describe in Chapter 5, Congress has blurred the legal difference between these two types of Public Information, but we will discuss how the distinction between them will continue to be of significance to preservationists.

It is worth noting that for more than 25 years, GPO has used the term IDP instead of Public Information in its policies that refer to the scope of what GPO and FDLP preserve. As early as 1995, GPO tentatively equated IDPs with "government publications" in its *Transition Plan* (GPO, 1995b). GPO's 2009 *Strategic Plan* defined the scope of FDLP as including all Federal information dissemination products published on an agency's website (GPO, 2009, p. 13). And subsequent policies on web-harvesting (GPO, 2016f), the scope of both the Cataloging program and the FDLP (GPO Superintendent Of Documents, 2019a), access to collections (GPO, 2019), and printing and publishing (GPO, 2020c) all refer to IDPs. This makes sense as a matter of practicality since information transferred to GPO for preservation must be packaged for transfer and that package is called an IDP.

Over the years, GPO has proposed revising the law to replace "government publication" with "information dissemination products" (GPO, 1995b; US Congress. House. Committee on House Administration, 2017, p. v.2 p.66). GPO may, however, be changing its preferred term to Public Information. In its 2020 and 2023 proposals for revision of Title 44, GPO dropped all mention of IDPs and used "Public Information" to describe its programs and scope (GPO, 2020b; GPO, 2023e). And, in its promotion of what it calls "The National Collection," it does not use the term IDP at all but does repeatedly refer to "Public Information" to describe both the scope of the collection and its services (FDLP, 2021b). This also makes sense because GPO's definition of the "National Collection" is not the description of an actual collection, but a description of the universe of information from which a collection could be built. Thus, GPO is using the term Public Information in the same way we are: as the universe of all government information released to the public, whether packaged into IDPs or not, whether distributed or not, whether preserved or not.

As noted above, these may not be perfect definitions, but they are significant because of their legal weight. We will, therefore, use these terms throughout this book and capitalize them to call attention to their specific meaning as described in this chapter.

CHAPTER 4
What is Preservation?

The meaning of preservation has changed significantly from the paper era to the digital age. In their textbook on the foundations of library and information science, Richard Rubin and Rachel Rubin note that paper-era preservation was primarily designed to protect an item from loss or damage, and use of preserved items might actually threaten their preservation. "In contrast," they write, "preserving information in a digital format is intended to increase access" (Rubin & Rubin, 2020).

Definitions of the word "preservation" often reflect the old, paper-era view or are just too vague to be useful to digital preservationists. In its loosest definition (and the first in the *Oxford English Dictionary*), it is simply the action of preventing some *thing* from decaying or being damaged or destroyed. This is both incomplete and misleading as a guide to planning for the preservation of digital information. In a 2022 report, Ithaka S+R found that the term "preservation" had become so devalued that it had lost its meaning (Rieger et al., 2022).

As Ken Thibodeau, the director of the Electronic Records Archives Program at NARA, said, the ideal preservation system would be a neutral communications channel for transmitting information to the future (Thibodeau, 2002). What would that look like?

What Preservation Is

The National Digital Information Infrastructure and Preservation Program (NDIIPP) defined digital preservation as "the active management of digital content over time to ensure ongoing access" (National Digital Information Infrastructure and Preservation Program [US], 2018). This is a good, succinct definition that is worth examining in more detail.

In this book, we rely heavily on the ISO standard known as OAIS (the *Open archival information system [OAIS] - Reference model*) as a guide to preservation vocabulary, functionality, and responsibilities. The OAIS definition of an "OAIS Archive" somewhat parallels, but expands, the NDIIP definition of preservation.

> An OAIS Archive is one that intends to preserve information for access and use by a Designated Community. (ISO, 2012b)

The scope of OAIS is not limited to digital information. It is applicable to the preservation of everything from moon rocks to books, to manuscripts, to digital objects. It is a "reference model," which means it does not specify any specific way to implement preservation. It does not specify or require any specific media or data formats or metadata standards, or hardware or software or institutional organization. It does, though, provide a model that preservationists can use to describe, design, discuss, and evaluate preservation policies and actions. It does this by defining the functions and responsibilities of an archive. We devote an entire chapter to OAIS (Chapter 15), but for our definition of preservation, we will present several of its essential concepts here. [The numbers in brackets refer to the numbered sections of OAIS.]

Three Key OAIS Concepts

There are three OAIS concepts that directly relate to defining preservation.

The Information, Not the Institution

The standard defines an archive as an Open Archival Information System and spells out the obligations of organizations that accept responsibility to preserve information. But the standard makes it clear that the organization is secondary to the information it preserves.

The information being maintained has been deemed to need Long Term Preservation, even if the OAIS itself is not permanent. [1.1]

Thus, preservation is about the information being preserved, not the institution that preserves it.

The Designated Community

The concept of the "Designated Community" is fundamental to OAIS. The term is used in every section of the standard and is mentioned more than 75 times. One of the six mandatory responsibilities of OAIS is for the Archive to identify its Designated Community or Communities. The standard is very clear: Information is primary, and the Archive is preserving the information for use by its Designated Community.

A Designated Community is:

An identified group of potential Consumers who should be able to understand a particular set of information. The Designated Community may be composed of multiple user communities. [1.7]

Thus, preservation for an OAIS Archive is not about just preventing something from being damaged or destroyed or lost; it is about making sure that the Information can be used by people.

The Long Term

Preserving information for the long term is a key aspect of OAIS. Although it is a general model that can be used for ensuring the preservation of all kinds of objects (including non-digital objects), it is scrupulous in covering the needs of digital information. It specifically addresses "permanent, or indefinite Long Term, preservation of digital information." The concept of the Long Term is particularly complex for digital objects, making its definition particularly important for preservation planning. That definition includes changing technologies, but also, notably, changing user communities. And the length of time of the "long term" is defined as being "an indefinite period of time" and into "the indefinite future."

> Long Term: A period of time long enough for there to be concern about the impacts of changing technologies, including support for new media and data formats, and of a changing user community, on the information being held in a repository. This period extends into the indefinite future. [1.7]

Thus, preservation is about access and use of information in the future.

The Five Guarantees

OAIS provides lots of technical details, but its essence is simple. As quoted above, "An OAIS Archive is one that intends to preserve information for access and use by a Designated Community." We can see what that means to the practicing preservationist by teasing out of the standard what it says about making the preserved information usable by a Designated Community.

In describing high-level OAIS concepts and functions, we see OAIS as making five guarantees about the information preserved. Information must be:
- not just preserved, but discoverable [2.2.2]
- not just discoverable, but deliverable [2.3.3]
- not just deliverable as bits, but readable [2.2.1]
- not just readable, but understandable [2.2.1]
- not just understandable, but usable [4.1.1.5]

This tells us that preservation is not passive, but functional. Our list of guarantees is comparable to the 2016 principles for scientific data management and stewardship that "all research objects should be Findable, Accessible, Interoperable and Reusable (FAIR)" (Wilkinson et al., 2016). As we mentioned in Chapter 1, the value of government information is only a potential value until it is activated by being usable and used. And usability is only the end point of a sequence of properties that includes discoverability, deliverability, readability, and understandability. If information sits in a "dark archive," unfindable and unusable, its potential value goes unrealized.

These are not new concepts. Linking information to people through the broad concept of "access" is a widely accepted standard for preservation. David Brunton, a supervisory IT specialist in the Library of Congress Repository Development Center, has said, "Digital preservation is access...in the future." His colleague, Ed Summers elaborated on this, pointing out that "access" begins now: "The underlying implication here is that if you are not providing meaningful access *in the present* to digital content, then you are not preserving it" (emphasis added) (Summers, 2013). The Council on Library and Information Resources made the same point in the title of an influential book of essays on stewardship of the cultural and intellectual resources entitled *Access in the Future Tense* (Greenstein, 2004).

We examine the concept of "access" in detail in Chapter 17, but for now the best articulation that we have seen of the practical connection between preservation and access and use is from prominent digital preservationist Paul Conway:

> For years, preservation simply meant collecting. The sheer act of pulling a collection of manuscripts from a barn, a basement, or a parking garage and placing it intact in a dry building with locks on the door fulfilled the fundamental preservation mandate of the institution. In this regard, preservation and access have been mutually exclusive activities often in constant tension.
>
> In the digital world, the concept of access is transformed from a convenient byproduct of the preservation process to its central motif. The content, structure, and integrity of the information object assume center stage; the ability of a machine to transport and display this information object

becomes an assumed end result of preservation action rather
than its primary goal. (Conway, 1996)

This brings us full circle, back to the beginning of this chapter and the primitive definition of preservation as just preventing something from decaying or being damaged or destroyed. In that understanding, preservation and access are "mutually exclusive activities often in constant tension" because the best way to prevent damage of something is to prevent people from handling it and using it. Such an understanding is no longer adequate because the concept of access has been transformed in the digital age "from a convenient byproduct of the preservation process to its central motif." Or, as Ithaka S+R determined, the goal of preservation is "to enable discovery, access, and use of content by designated user communities over time" (Rieger et al., 2022).

Definition of Digital Preservation

Bringing these concepts together, we suggest a working definition of digital preservation.

> Preservation means maintaining information for discovery,
> delivery, readability, understandability and usability by a
> Designated Community for the long-term.

When we use the term "digital preservation" in this book, this is what we mean. It is more than storing digital stuff. Storing digital stuff is important, of course. We might say that ensuring that information is safely saved and protected from alteration or loss is the foundation on which preservation is built. But preservation is, following OAIS, explicitly about maintaining information for long-term functional use by people. In the digital age, access today does not guarantee access tomorrow, and storage without access is not preservation. In the digital age, preservation must be inextricably linked to access. In Part Three we elaborate on this brief definition by delving into the details of what terms like discovery, delivery, and usability mean in the digital age. We will also suggest that, for government information to be adequately preserved, preservationists must be able to select information comprehensively and acquire sufficient control over it to ensure its preservation.

PART TWO
Where Are We and How Did We Get Here?

To plan for the preservation of government information, preservationists need a clear understanding of what is being preserved and what is not. An overestimation of either can intimidate preservation efforts or lead to a false sense of accomplishment. Without an accurate understanding of the content that needs to be preserved, planning will be based on guesses, selection will be done in the dark, and a combination of unnecessary duplication and missed content will result in inefficiencies and preservation gaps. So, in Part Two, we examine what is being produced and what is being preserved.

A lot of good work is being done to save and store born-digital government Public Information. GPO has a certified trusted digital repository and its own web-harvesting program (GPO, 2021a). The Internet Archive crawls and stores great swaths of the web, including the government web. The End of Term Archive project has stored the state of the government web just before and just after each presidential inauguration since 2008. The National Archives uses the same criteria for preserving government digital records that it has successfully used for analog records for decades. The Library of Congress has an active web preservation plan that includes government web sites. Many government agencies are actively using the web to make their information directly available to the public and some of them keep historical information online indefinitely. Government data has its own website (data.gov) for helping people locate selected data. University libraries are actively preserving content needed by their faculty and students. Data archives like ICPSR are actively preserving government data. Specialized archives are preserving content in important areas that would otherwise be potentially neglected (ICPSR, 2025b). A robust private sector information industry profits from obtaining, saving, and redistributing government information. And yet...

With all this activity and the undisputed availability of an unimaginable amount of government-produced digital information, it would be easy to infer that digital government information is being adequately preserved. But the availability of information on government websites today does not guarantee access to that information in the future. And there is no way to accurately measure how complete or incomplete the current preservation activities are because there is no complete bibliographic record, and we have no reliable way of measuring either how much is being produced or how much is being preserved.

In Part Two, therefore, we want to go beyond any vague confidence that government information is being adequately preserved and any vague fear that there is so much to preserve it will have to be up to someone else to solve the preservation problem.

To do that we will examine the current state of preservation of digital government information. Our primary focus here is on born-digital Public Information that is posted on the web. Our goal is to give a realistic appraisal of where we are and how we got here so that we can begin to see where we need to go and how we can get there.

We begin with an overview of the laws of preservation (Chapter 5). Then we use the 2020 End of Term web harvest to characterize the extent of government Public Information on the web (Chapter 6). Next, we analyze the content of the GOVINFO digital repository (Chapter 7). Then we review web harvesting of the government web (Chapters 8-12). We conclude Part Two by enumerating six preservation gaps (Chapter 13).

CHAPTER 5
The Laws of Preservation

Given that preserving government information is a widely accepted principle (Chapter 1), one might reasonably expect that the obstacles to doing so would be technical or budgetary (i.e., "We'd like to do that, but we don't have the tools—or the money"). There are indeed technical hurdles to preserving digital information successfully. And the costs of ongoing, active digital preservation and access are greater than the cost of passive paper preservation. But it is laws that enable funding or, in the worst case, impede action. Laws are a key component of the infrastructure of preservation. Laws determine what can be preserved and what is available to be preserved and by whom. If the laws are inadequate or badly enforced, preservation will be inadequate regardless of principles or funding or technologies.

Introduction

There is very little in the *United States Code* that explicitly requires preservation of government information. The laws governing information policy and preservation are complex, arcane, sometimes vague, and often contradictory. They were, for the most part, written in the paper era and were designed to accommodate paper-based information. Accommodation by these laws to digital information has mostly been tacked on after the fact, often by simply modifying definitions of information to include digital formats, but without either taking into account the inherent differences between analog and digital information or attempting to develop a preservation infrastructure.

The National Archives and Records Administration (NARA) has published a book of almost 200 pages of "Basic Laws and Authorities" that affect NARA with a note that additional materials can be found on the Archives' website (NARA Office of General Counsel, 2016). The scope of these laws, however, is not exclusively or even primarily about preservation—most are about records management.

Similarly, the scope of the laws and regulations that define government information policy focus more on the collection, creation, and dissemination of information than on its preservation. They affect preservation but say little about it. A lexicon of information terms used in US laws, regulations, and policies that permeate every aspect of the federal information system is in its sixth edition and is over 400 pages long (Maret, 2016). A 1996 Congressional Research Service (CRS) study found and reprinted more than 300 pages of laws, regulations and policies relevant to Government information dissemination in more than 250 sections of the *US Code* (Griffith & Relyea, 1996). (A list of those laws was published as a 10-page attachment to a 1996 study of the FDLP [GPO, 1996b].) In 2001, the National Commission On Libraries And Information Science (NCLIS) updated CRS's compilation with a 40-page list of major legislation enacted between 1995 and mid-2000 (US National Commission On Libraries And Information Science, 2001). In just that five-year period, NCLIS found 52 public laws enacted affecting more than 650 sections of Statutes. Significantly, NCLIS found the laws confusing and contradictory and noted that not all agencies recognized or understood their responsibilities under the laws. Other studies, including a 2018 Library of Congress survey, have also found that agencies are often unaware of the law or their responsibilities (US Library of Congress, Federal Research Division, 2018).

Some agencies (e.g., the Energy Information Agency [EIA], the Environmental Protection Agency [EPA], the Bureau of Labor Statistics [BLS], and others) have mission statements that include keeping their older Public Information accessible on their websites. But there is rarely, if ever, an explicit legislative requirement for agencies to preserve that information. Indeed, while we were writing this chapter, the EPA announced plans to retire its online archive (Calma, 2022; US Environmental Protection Agency, 2025). Agencies' policies are subject to interpretation, modification over time, and funding shortfalls. The *US*

Code also explicitly permits agencies to charge fees for access to some information. See, for example, the Administrative Procedures for public information that apply to government agencies (5 USC 552) and the *GPO Access Act* (44 USC 41). See also the story of access to court opinions in Chapter 7.

There has been substantial progress in mandating preservation in some specific areas. For example, after a 1990 GAO report found serious problems with the National Aeronautics and Space Administration's (NASA) archiving of "irreplaceable" space science data, Congress passed Public Law 102-555, which set up the National Satellite Land Remote Sensing Data Archive (NSLRSDA) (US General Accounting Office, 1990; Bleakly, 2002). NASA now has elaborate procedures for preserving its scientific data and making it available (Behnke et al., 2019). Its Records Retention Schedule that treats scientific data as permanent Records has also been approved by NARA (NASA, 2023).

Despite the complexity and confusion of information laws, Congress does address preservation of Public Information. It has assigned preservation not to the agencies that create the information but to three preservation agencies: the Government Publishing Office (GPO), the National Archives and Records Administration (NARA), and the Library of Congress (LC). These three agencies have explicit legal preservation mandates. The three differ from other agencies in that they manage information from other agencies. These preservation offices have to acquire that information—or acquire control of the information—in order to preserve it and ensure its availability to the public.

In this chapter, we will examine laws governing the preservation of federal Public Information by these three agencies in the paper era and the digital age, with particular attention to information posted on government websites.

GPO

The Paper Era

Most of what we now call Public Information was published over the years as paper and ink, mostly by the "Government Printing Office" (GPO). Those publications were sent to Federal Depository Libraries (FDLs), which provided access in the at-the-time-traditional way of local libraries open to the public; all FDLs are required to make deposited publications "available for the free use of the general public" (44 USC 1911). Those publications were preserved as a byproduct of the process of multiple copies being distributed to and legally retained in multiple depositories. Lorcan Dempsey of OCLC, writing about academic libraries, described this as the "print logic" of libraries, noting that preservation was a "side-effect of the redundancy of the print distribution model" (Dempsey, 2017). This was generally true even before 1962 when the law was amended to require a subset of FDLP libraries, called "Regional Depositories," to retain government publications permanently (44 USC 1912). Laws and regulations and practices converged into a clear path—an infrastructure—of long-term preservation of and access to published government Public Information. It was taken for granted in the print era that, once government information was released to the public, it could not be withdrawn or altered and would not be lost.

The Preservation Gap

The system was not perfect. Most of the Public Information of the judicial branch was never formally published at all and, so, never made its way to GPO or depository libraries. And when an executive branch agency simply failed to send a publication to GPO for printing, that publication would not be cataloged or sent to depository libraries (Jacobs, 2017). There were many reasons why executive branch agencies would withhold their publications from GPO, but the problem worsened drastically after a 1983 Supreme Court decision that had the effect of giving executive branch agencies permission to refuse to send publications to GPO (*Immigration and Naturalization Service v. Chadha*). This damaged preservation by creating an unknown number of unreported (often called "fugitive") publications (Jacobs, 2021) outside the existing preservation infrastructure (Petersen, 2017). Executive agency decisions subsequent to *Chadha* further exacerbated the problem (GPO, 2016b).

These problems created a preservation gap in the print-era infrastructure. The law gives GPO the authority to preserve publications it publishes, but it does not give GPO the authority to enforce its acquisition of executive branch publications. The amount of executive branch Public Information published by GPO dropped drastically in the 1980s and 1990s. It seems that executive branch agencies were either unaware of the preservation infrastructure or did not want to pay for preservation copies or, perhaps, just couldn't be bothered with the paperwork (GPO. Office of Inspector General, 2017; US Library of Congress, Federal Research Division, 2018). Whatever the reasons, the result was that massive numbers of publications no longer had a working path to preservation.

Although the preservation gap was large in the paper era, the digital age compounded the problem exponentially. GPO recognized the gap between laws and practice as early as the mid-1990s (Baldwin, 1997). The preservation gap for executive branch publications expanded from, perhaps 50 percent in the late print era (Baldwin, 2003) to near 100 percent in the digital age. With the advent of digital publishing direct to the web, by 2011, 97 percent of new federal government publications were not published as paper and ink but were simply posted on government web sites (GPO, 2011a). The 2018 Library of Congress survey of agencies found that most were publishing 100 percent of their Public Information digitally (US Library of Congress, Federal Research Division, 2018). Almost none of these was deposited in any form with depository libraries.

The GPO Access Act

Congress's first attempt to deal with the shift to digital publishing was to direct GPO to provide online access to some federal electronic information and to store that information in an "electronic storage facility." This law, the *Government Printing Office Electronic Information Access Enhancement Act of 1993*, often referred to as the *GPO Access Act*, resulted in the creation of *GPO Access* (access.gpo.gov, gpoaccess.gov), an online service that became the *Federal Digital System* (gpo.gov/fdsys/) in 2012 and *govinfo* (govinfo.gov) in 2018. The title of the act reflects the focus of the legislation on "access" rather than preservation. The purpose of the act is defined in its first sentence: "...a means of enhancing electronic public access..." and it directed GPO to create "an electronic directory of Federal electronic information." The law was codified into Chapter 41 of Title 44 of the *US Code*, separate from the depository library program in Chapter 19, and it differs significantly from Chapter 19 in several ways. Instead of requiring free access for the public, it allows GPO to charge for access. Instead of requiring distribution of digital publications to depository libraries and requiring them to retain them, it simply directs GPO to store publications without any specific requirement for long-term or permanent retention or preservation (US General Accounting Office, 2001, p. 10). Instead of mandating preservation of all government Public Information, its scope is left open for interpretation.

The act specifies only two publications for the storage facility and online access, the *Congressional Record* and the *Federal Register*. It does, however, allow GPO to provide access and storage to "other appropriate publications distributed by the Superintendent of Documents." That permission echoes the language of the depository library program in Chapter 19, which is how GPO has interpreted the law (GPO Superintendent of Documents, 2019a). But this does not address the fact that most Public Information (i.e., government information on the web) is *not* distributed by the Superintendent of

Documents. The act allows the Superintendent of Documents to include agency information in the storage facility ("to the extent practicable") if the agency requests it, but it does not require agencies to request such storage and, once again, does not give GPO any enforcement authority.

The law introduces a new term, "Federal electronic information," and defines it as "Federal public information stored electronically." This can be interpreted as broadening the potential scope of preservation to include all federal government digital Public Information. This is a great improvement over the Chapter 19 scope of government "publications" published as "individual documents." But, without giving GPO any authority to enforce the acquisition of Public Information, the law left the preservation gap in place.

The best that can be said of the law is that it gave GPO the leeway to infer responsibility for preservation without providing any requirements that it do so and without giving it the authority to enforce any of its preservation policies.

GPO and others have implicitly recognized these weaknesses of the law by proposing changes to the sections of Title 44 that deal with the depository program and GPO's online access service. Among those proposals, GPO has recommended incorporating Chapter 41 into Chapter 19 (GPO, 2017). None of those proposals has passed, however, leaving GPO without the authority to acquire control of Public Information for preservation and without a clear legal mandate to preserve web-based Public Information.

Implementing the Law

The vagueness of the law has given GPO the ability to change policies over time. For example, GPO initially charged the public for use of *GPO Access*. It did give free access to users of Federal Depository Libraries, however, but it limited FDLs free access to 4, and later 10, users at a time. GPO stopped charging for access less than two years later after some depository libraries developed ways to provide the same content online for free (GPO, 1994; Relyea, 2004). GPO spent years trying to use web-harvesting to collect only web content that conformed to the Chapter 19 definition of "government publications" and reject "out of scope" content. Then, in 2008, it simply decided to interpret the law differently and accepted content published on an agency's website as in scope (GPO Superintendent of Documents, 2019a) (see Chapter 9). The laws that allow GPO to charge fees for access and that define the scope of preservation remain unchanged, but GPO's policies implementing the law have changed. This could be seen as an advantage to GPO and preservation: policies can adapt to perform better. But flexibility can go both ways. Without a clear legal mandate, it is easy for political considerations, bureaucratic decisions, and budgetary priorities to weaken preservation policies. It makes it easier for GPO to request smaller budgets and proclaim cost savings even if that limits what it preserves. And it makes it easy for Congress to deny funding for ambitious projects.

Despite the weaknesses of the law, GPO has interpreted it as a mandate to preserve digital government information. As evidence that GPO takes its self-appointed role of preservation seriously, it began working toward building an OAIS-compliant digital archive in 2004 and, 14 years later, obtained Trusted Digital Repository certification for its repository (GPO, 2004b; PTAB Primary Trustworthy Digital Repository Authorisation Body Ltd, 2018).

The Digital Preservation Gap

Despite GPO's good intentions, the weaknesses of the *GPO Access Act* suggest that the preservation gap, particularly for executive branch publications, will persist and grow as digital publishing increases. To test this, we compared the number of PDF files harvested by the 2020 End of Term web crawl from executive branch domains with the number of PDF files in executive branch collections in the GOVINFO repository as of early 2022. The 2020 EOT crawl gathered about two million PDFs from the executive branch. In contrast, during the almost 28 years of its existence, the GPO storage system had acquired fewer than 52,000 born-digital PDFs from the executive branch—about 2 percent of what the EOT crawl gathered in a couple of months.

Our analysis of the 2020 EOT harvest (see Chapter 6) reveals that, of the three branches, the executive is, by far, the biggest publisher of Public Information on the open government web. Specifically, it accounts for almost 3 times the publication output of Congress and more than 10 times the publication output of the judicial branch. In contrast, our analysis of the content of the GOVINFO repository (Chapter 7) reveals that only 2 percent of its holdings are from the executive branch. Clearly, the preservation gap still exists in spite of the *GPO Access Act*.

Summary

It is safe to conclude that, due to the limitations of the law and in spite of GPO's good intentions, GPO has been unable to acquire, provide access to, create a directory of, or preserve most government published Public Information. Almost 30 years after passage of the *GPO Access Act*, the great bulk of federal executive branch Public Information remains absent from *govinfo.gov*. The law failed to create a legal path for Public Information on the web to be preserved by GPO or FDLP libraries and failed to close the preservation gap.

GPO has, however, experimented with web harvesting as a supplement to its traditional Title 44 acquisitions policies. We examine its web harvesting history in Chapter 9.

NARA

The Paper Era and the Digital Shift

In the paper era, Public Information Records and published Public Information were, for the most part, easily distinguished and were handled separately and differently (by NARA and GPO, respectively). By law and practice, Records were treated as archival material, which typically got less detailed indexing (only collection- or agency-level metadata) and limited or restricted public access at a small number of NARA facilities. In contrast, "publications" were distributed to more than 1,000 libraries for access by "the general public" (44 USC 1911).

With the digital shift, it became less clear which laws applied to documents on the web and therefore to almost all published Public Information. Should they be treated as Records under the *Federal Records Act* (FRA) for potential transfer to the National Archives, or Publications to be distributed to FDLP libraries under Title 44 of the *US Code*? Without a clear legal path to preservation, published Public Information was falling through the cracks. The law did not specify who was legally responsible for preserving digitally published Public Information, and there was no infrastructure in place to ensure that it would not be withdrawn, altered, or lost.

The National Archives was an obvious candidate for addressing long-term preservation of digital information, including published Public Information, and there were good reasons to imagine that NARA could take on this role.

First, there is the fact that NARA has a history of preserving paper-based published Public Information. A 1968 law (Pub. L. 90-620) requires GPO to deposit several titles with NARA. And, in 1972, when GPO found itself running out of space, it transferred its entire Public Documents Library (dating back to 1895) of almost two million volumes and 76,000 maps to NARA (Barnum, 2011). GPO has continued to transfer printed publications to NARA in multiyear blocks (NARA, 2020a). This is a far from complete collection even of documents deposited with Federal Depository Libraries. NARA describes the Record Group for this collection as consisting of only one-half to two-thirds of all government publications (NARA, 2016). Although these documents are preserved by and available from NARA, access to them is limited. NARA describes it as a "resource of last resort" for researchers who cannot locate copies in depository libraries (NARA, 2021; Holterhoff, 1990b). (In 2004, GPO proposed that NARA move the collection to "storage" and make it a "dark archive" with even less public access [GPO, 2004a].) Nevertheless, it does provide an indication of NARA's ability and willingness to provide long-term storage for (if not easy access to) published Public Information, at least in paper form.

Second, NARA has a decades-long history of preserving digital information. At least as early as 1939, when Records were defined to include punch cards and tabulation sheets (53 Stat 1219), NARA has dealt with computerized information and, at least since the 1950s, NARA has identified, preserved, and provided access to computerized information (Hull, 1998). It established a "data archives staff" in 1969 (Committee on the Records of Government et al., 1985), giving it more than 50 years of experience archiving electronic records (Henry, 1998). NARA created its own Center for Electronic Records in 1988 (Rowe, 1992) and issued rules for Electronic Records Management shortly after that (NARA, 1990). Congress has expanded NARA's scope to include digital information by adjusting the definition of "Records" over the years to include new types of materials, platforms, and the physical media used to store information (Stuessy, 2019).

Records or Publications?

In spite of this history, as recently as the early 2000s, no law specifically defined what we now call published Public Information on the web as either Records or as Government Publications, nor was there any direct legal path for this information to go to NARA. The result, as one Senate Report described it, was that agencies were having "particular difficulty complying with the *Federal Records Act* (FRA) with respect to electronic records, including records posted on the Internet" (Senate Committee on Governmental Affairs, 2002).

Congress finally addressed this with the *E-government Act of 2002*. At that time, government use of the web was still in its early stages, and agencies lacked direction from Congress on how to use the web, and the *E-Government Act* was passed mostly to address this issue. It was a hopeful law that seems almost quaint in retrospect. It had to direct the federal courts to establish their own websites and federal regulatory agencies to put their dockets on the web. It was intended to improve online access to agency information, and it directed the creation of a "federal Internet portal" that would "improve public access to government information and services." It also directed the creation of a "directory of government websites organized by subject matter...linked to the federal Internet portal."

It is notable that, nine years after passage of the *GPO Access Act,* which directed GPO to create an "electronic directory of Federal electronic information," Congress found it necessary to take another try at creating a directory of electronic Public Information and at making government information more easily accessible on the web.

But Congress did include preservation in one section of the law (Sec. 207 "Accessibility, Usability, And Preservation Of Government Information"). The law addressed the problem of web-based information falling through the cracks (i.e., Records or Publications?) by defining the content on government websites as Records and by directing NARA and government agencies to treat them as such. It created the Interagency Committee on Government Information (ICGI), with one part of its charge being to recommend policies and procedures to NARA that would ensure that government information on the Internet would be treated as Records under the *FRA.*

The Senate report on the bill described Congressional intent: It would "improve preservation of, and public access to, electronic information by achieving greater compliance with the *Federal Records Act.*" The ICGI would recommend the adoption of "standards to achieve greater compliance with the *Federal Records Act.*" The *FRA* applies only to executive branch agencies and courts and does not apply to Congress, the president, or the Supreme Court (Ginsberg, 2015). Executive branch agencies produce the great bulk of the government's published Public

Information (see Chapter 6), so the law's scope is massive, but not universal.

NARA's Guidance

The *E-Government Act* answered one question, but it created a new one. It answered the question of whether published Public Information should be treated as Records (under the *Federal Records Act* provisions of Title 44 of the US Code) or as Publications (under Chapter 19 of Title 44). It did so by simply asserting that it should be treated as Records. But this created new questions for NARA and agencies. How would they apply the old, paper-era laws of records management and preservation to a class of information (Publications) for which those laws had not been designed?

Those paper-era laws do not require NARA to preserve all Records; they require agencies to plan for their disposition. "Disposition" can include keeping the records, turning them over to NARA for long-term preservation, or discarding them. In practice, only 1 to 3 percent of federal Records are scheduled for permanent preservation (NARA, 2020b).

The decision as to whether an agency's Records will be preserved or discarded begins with the agency itself, which develops a records management plan (called a "Records Schedule"), which it submits to NARA for approval. Agencies develop their plans based on the law, the regulations, and the guidance provided by NARA.

In September of 2004, before the Interagency Committee issued its preservation recommendations, NARA had already issued to agencies its first guidance on web-based information. That first guidance asserted that "Web content records are a priority electronic records format" for NARA, and it described the technologies for transferring web content to NARA.

Three months later, the ICGI issued its recommendations on treating web content as Records (Interagency Committee on Government Information Electronic Records Policy Working Group, 2004). Two months after that, NARA issued new, more detailed guidance, acknowledging explicitly that the *Federal Records Act*, which charges NARA with preserving historical records for posterity, "applies to all agency records, including web records" (NARA, 2005d).

The *Federal Records Act* allows agencies to discard records that are "not needed by it in the transaction of its current business." In the paper era, this meant that most Records were discarded because they were unpublished internal records that were no longer needed for "current business." This makes sense because the information being discarded under this guideline had never been made available to the public and probably had only temporary value to the agency.

Published Public Information, however, possesses two inherent characteristics that unpublished Records do not. First, by the act of publishing information, an agency has already identified that information as having public value. Second, the publication of information is an act of government and the Publication itself is a record of that act. These characteristics alone suggest that most published Public Information is at very least suitable for preservation. That was, indeed, the default in the paper era when the law (Title 44) treated Publications and Records differently.

By defining web-based publications as Records, however, the *E-Government Act* imposed the Records criteria for preservation on Publications. In addition to allowing agencies to discard records that are not needed for its current business, the *Federal Records Act* also allows agencies to discard materials "that do not appear to have sufficient administrative, legal, research, or other value to warrant their further preservation by the Government" (44 USC 3303). In other words, in order to justify preserving web-content Records, an agency would have to find "administrative, legal, research, or other value" in that content. Although preservationists might agree with us that the mere act of publishing the information suggests web-based content has one or more of those values, agencies and NARA might disagree, and it is the agencies and NARA that have the decision-making power under this law.

If they didn't disagree in principle, they might disagree as a practical matter since neither individual agencies nor NARA have been funded or staffed to provide long-term access to the huge volume of information that agencies publish on the web.

In its 2005 guidance document, NARA took a practical approach. That guidance to agencies states that most web records do not warrant permanent retention and should be scheduled for discard. Web records, NARA said, should be discarded "as soon as records are no longer needed" to conduct agency business. The guidance also includes helpful examples for the wording of records schedules for web records, and every example is for the destruction of records. These include destruction of Inspector General reports, Privacy Act reports, and "monthly snapshots of web content." The Guidance provides no special instructions for the preservation of published Public Information.

After reiterating that "most web records do not warrant permanent retention and should be scheduled for disposal," NARA's guidance does mention one reason an agency might transfer web content to NARA for long-term preservation. That is if NARA, not the agency, finds websites to have historical value:

> In instances where *NARA determines* that a site or portions of
> a site has long-term historical value, NARA will work with
> the creating agency to develop procedures to preserve the
> records and provide for their transfer to the National
> Archives. [emphasis added] (NARA, 2005d)

This leaves web-based Records in an extremely tenuous, almost catch-22 situation. Although the agency that produces the information could choose to keep its information online forever, few have either the legal mandate or the resources to do so. Although NARA could override an agency's decision to discard web content, it also lacks the resources for such a massive preservation project on its own (see Chapter 10). Below, we examine how this has worked out in practice.

Agency Implementation of the Law

The practical result of the law and NARA's guidance and policies is that most if not all agencies describe their websites as "temporary" and the final disposition of those website "Records" will be to discard them. NARA is approving those schedules.

This outcome is consistent with NARA's advice to agencies, which is itself clear and consistent. NARA emphasizes that preservation should depend primarily on the "business needs" of the agency. Thus, the instructions are to destroy records when they are superseded or obsolete. Similarly, NARA tells agencies to identify a "recordkeeping copy" of documents that warrant preservation and discard other copies. Agencies typically identify documents on their websites as copies that can be discarded because the agency's other information systems contain the official recordkeeping copy (NARA, 2019b). A survey of agencies by NARA in 2017 found that "very few agencies (28%) transfer permanent web records to NARA" (NARA, 2018).

Examples of how agencies have applied NARA's guidance abound, and they often use the wording in the guidance word-for-word. For example, the Nuclear Waste Technical Review Board designates web content for immediate destruction when "no longer needed for conducting business." The board is responsible for scientific and technical reviews of the management and disposal of high-level radioactive waste and spent nuclear fuel, and it provides findings and recommendations to Congress, the Secretary of Energy, and "the interested public." The web content scheduled for destruction includes year-end reports, files related to its mission, press releases, budget justifications, and the agency's strategic plan (Nuclear Waste Technical Review Board, 2014).

Similarly, the Securities and Exchange Commission (SEC) designates its web content as "temporary" and says that it should be destroyed when no longer needed because its web content "is maintained offline" with its own records schedule. The SEC notes that content is "removed from the Web site when it is superseded, obsolete or no longer needed for SEC business" (Securities and Exchange Commission, 2019).

The Consumer Financial Protection Bureau (CFPB) uses almost identical wording to designate its web content as temporary information that can be destroyed because "[t]he official records of published Web Content are stored and managed offline by the authoring Division/Office in accordance with NARA approved records disposition schedules" (Consumer Financial Protection Bureau, 2019).

The Grants.gov website is designated as temporary because the information on the site is posted there not by Health and Human Services, which runs the site, but by other agencies that presumably manage that content "as records in their own records-management environments." NARA notes that "[Grants.gov] has the unique and sole business purpose of making grant information available to the public for a relatively short period of time" (NARA, 2021b).

A few agencies have scheduled at least some of their web content for public preservation. In each case that we found, however, the agency only did so when it could identify web content that was *not* in the agency's internal recordkeeping system. The State Department, for example, scheduled its website (state.gov) for "temporary" disposition with pages to be deleted as they are superseded or are no longer needed for reference because web content contains duplicate information maintained in other department recordkeeping systems. But the same State Department Records Schedule (US Department of State, 2021) classifies its "DipNote Blog Page" with "permanent" disposition with transfer to the National Archives at the end of each calendar year, presumably because it contains content that is not in the department's recordkeeping system.

Problems of Treating Published Information as Records

By defining published Public Information as "Records," the *E-Government Act* creates several preservation problems. Information published on websites for the public is inherently different from unpublished government Records. It is, by definition, "Public Information," a subset of Records (see Chapter 1), and therefore should be treated differently. The paper-era law (44 USC 1911-1912) designated all published "Government Publications" for permanent preservation in Federal Depository Libraries, while NARA has traditionally preserved only 1 to 3 percent of unpublished Records (NARA, 2020b). As noted earlier, there are significant reasons for these different preservation approaches, and those reasons have not changed in the digital age.

These raise problems of selection and access.

With regards to selection, NARA's policies deprecate selection of published government information. Its policy on "Government Publications and Library Materials" including "electronic" information (NARA, 2019a) explicitly says that "[i]n general, published materials are considered to be 'non-record' material and NOT included among the records holdings of NARA. **NARA is a source of last resort for U.S. Government Publications**" (emphasis and capitalization in original). In addition, NARA's "Guidance on Managing Web Records" says explicitly that "web content" should be destroyed "when superseded, obsolete, or no longer needed for the conduct of agency business" (NARA, 2005d). Such information is essential to making government actions transparent and keeping government accountable to citizens. It is vital to those who wish to understand the evolution of government policies that "superseded" information (updates, alterations, deletions, versions) be preserved.

With regard to accessibility of information preserved, one must ask if the procedures NARA uses for making unpublished Records accessible will be adequate for published Public Information. Will, for example, the "recordkeeping" copies of content that an agency published on the web be packaged for public discovery, distribution, and use? Or will that information be buried in databases and entangled with other unpublished Records that were not designed for public release? Will NARA have the policies or budgets to provide robust accessibility to individual published documents, or will they only provide broad access to bundles of record groups?

These questions deserve examination, and a successful Digital Preservation Infrastructure should resolve them.

Summary

The story of NARA's preservation of websites is complex, but two things are clear. First, the *E-government Act* changed the preservation standard for published Public Information from the retain-everything-published-by-GPO standard of 44 USC 1911-1912 to the retain-only-1-to-3-percent standard of traditional records management. Second, agencies are scheduling web content for discarding.

But the role of NARA in preserving websites is a bit more complicated than that since NARA has used web harvesting to preserve websites. We analyze NARA's web harvesting in Chapter 10.

Library of Congress

The Paper Era

The Library of Congress has a long history of collecting, preserving, and providing access to the federal government's published Public Information. Some of this history is driven by laws (some as narrow as they are specific) that require deposit of congressional publications with the Library. Title 2 of the *US Code*, for example, requires deposit of "twenty-five copies of the public Journals of the Senate and House" and "two copies of the journals and documents and...each book printed by either House of Congress" (2 USC 145-146). A separate requirement in Title 44 requires "bound" copies of specific documents (such as bills and resolutions and House and Senate documents and reports) to be furnished to the Library and deposit of up to 25 copies of "all federal publications printed under the authority of law" (44 USC 1718).

The implication of these laws is that the library will keep these deposited publications available for use, i.e., preserve them. Sections 131 and 141b of Title 2 briefly but explicitly require preservation of collections. Preservation of information is, indeed, an inherent part of the Library's mission. LC has a broad general mandate as the nation's library and the largest library in the world to "collect, preserve, and provide access" (US Library of Congress, 2022a). In this context, LC has created a Preservation Directorate that is responsible for managing collections and ensuring long-term access to them (US Library of Congress, 2025).

In addition to these legal mandates, as the library "of Congress," LC has a long and special relationship to government information by and for Congress. LC has been a selective FDLP depository since 1977 and has a collection that stretches back to 1933 of more than one million volumes (US Library of Congress, 2022b; GPO, 2025c). Both the Library and its Law Library are part of GPO's "preservation stewards" program to selectively preserve print publications (US Library of Congress, 2020b).

The Library's holdings of federal government public information extend far beyond its separately housed depository collection. The library describes these holdings as comprehensive and unique in breadth and depth. Its holdings stretch back in time to the founding of the republic and include publications that were not widely distributed, and, in some cases, the Library may hold the only available copy. Because the Library's extensive holdings of federal government publications are housed in the depository collection, and in the general collection, and in other special collections, they warrant their own finding aids, which provide detailed listings of holdings including commercial finding aids and reprints (Sansobrino et al., 1993; Sweeney, 1992).

The laws requiring deposit of printed federal government publications, the Library's commitment to serving Congress, its aggressive acquisitions of government publications, and the Library's mandate to preserve its collections provided a solid legal basis for preservation of federal government Public Information in the print era.

The Digital Age

But the print-era laws are tied directly to print publications and the budgets for print publishing, so they have little or no explicit applicability to the same Public Information when it is published directly to the web. Wording about "printing" and "publishing" and "binding" permeates those paper-era laws and makes it difficult to extend their scope to digital publishing, much less to website content when policy interpretations of terms such as "deposit," "delivery," and "distributing" have been used to prohibit rather than enforce deposit or delivery of digital content to FDLP libraries (see Appendix B).

While these laws do not provide specific mandates for digital preservation that are analogous to the print-era mandates, the laws governing what LC acquires and preserves are more open-ended than those that govern GPO and NARA. Rather than limiting LC's preservation responsibilities, the law gives LC the flexibility to seek out and acquire information that needs to be preserved. This legal flexibility has allowed LC to acquire large quantities of digital government information, mostly through web harvesting. As for the preservation of digital information that it acquires, the Library has developed a Digital Strategy (US Library of Congress, 2019) of preserving digital collections including their metadata and their utility and supporting "trustworthy long-term stewardship."

Despite the lack of specifically digital legal preservation requirements, the Library has an active, aggressive program for acquiring and providing long-term access to at least some government websites. We describe LC's web harvesting activities and collections in Chapter 11.

Summary

In the print era, the Library was largely able to rely on federal agencies delivering their information to the Library for preservation because they were required to do so by law. In the digital age, when the law failed to require agencies to deposit their born-digital content with the Library, the

Library did not stop collecting; it switched tactics. It developed policies for selection and acquired the materials (mostly through web harvesting) that it determined its user communities needed.

The policies and actions of the Library of Congress suggest an approach to planning successful digital preservation. By simply following a traditional library mission of selecting and acquiring information for its user communities (US Library of Congress, 2016), LC has found a practical way of preserving digital government information in the absence of specific laws requiring it to do so.

CHAPTER 6
Federal Government Public Information on the Web

In order to evaluate the current status of the preservation of government information and evaluate the effectiveness of current preservation tactics, we need to be able to compare what is being preserved to the universe of what needs to be preserved. We have to have a clear understanding of what is being produced. In this chapter, we provide an analysis of the government web by using data from the 2020 End of Term (EOT) crawl. Our goal is to use existing data to develop a preliminary profile of federal government information on the web.

Introduction

One of the biggest concerns of digital preservationists is the volatile nature of the web. It is well documented that websites come and go, that content is withdrawn and moved ("link rot"), and that content of web pages and documents change ("content drift"). The government web is not immune to these problems. An analysis by the Chesapeake Digital Preservation Group (Chesapeake Digital Preservation Group, 2013) found that link rot for dot-gov websites was as high as 51 percent. Indeed, whole government websites come and go. The first step to understanding the volatility of web-based content is simply to know what is being produced. Unfortunately, there is no reliable, consistent data that describe what is being produced.

Preservationists also worry about the sheer volume of information being published by the government. That volume is often expressed in broad terms of byte count or file counts. Indeed, web archives often characterize what they preserve this way, using the number of terabytes or petabytes of data they have stored or the number of URLs or files they have harvested. Those metrics can be so large as be intimidating, particularly to those new to government information or new to digital preservation. But what if byte count and file count give us an unrealistic understanding of the amount of government information to be preserved? Lavoie and Dempsey of OCLC in their well-known 2004 examination of the issues of digital preservation noted that expectations that the costs of digital preservation will be formidable may dampen the impulse to provide adequate long-term funding to digital preservation projects (Lavoie & Dempsey, 2004). In a study that the Center for Research Libraries commissioned for its 2014 collections forum "Leviathan: libraries and government information in the age of big data," Jacobs included a graph that compared the 160 million URLs harvested by the 2008 End of Term Crawl to the two to three million items estimated to be held in the Federal Depository Library Program (FDLP) (Jacobs, 2014). The implication of that graph was that the amount of government information produced digitally (perhaps in a single four-year period) was more than 50 times the amount of paper government information accumulated over almost 200 years by the FDLP. But is that a correct inference from those two numbers? What if we are choosing preservation strategies based on unrealistic or imprecise measures of the size of the government web?

Our knowledge of the government web is so limited that, for many years, even such a simple fact as the number of federal government websites was a matter of conjecture and speculation. As recently as 2003, the California Digital Library (CDL) had to use a literature review, interviews, and a test crawl of selected web sites in an attempt to analyze the scope and composition of the web-based government information (Cruse et al., 2003). Counts also tended to vary with the method of counting and when the counts were done. Quoting surveys of the Domain Name System (DNS) between 1992 and 2003, the CDL study found that the number of "dot-gov hosts" varied between 46,000 and 834,000. The report concluded that "the government domain is relatively large, volatile, and opaque" and "the demographic characteristics of the dot-gov domain are not uniformly apparent." In

2009, the General Services Administration refused a Freedom of Information Act request to release a list of registered dot-gov domains saying that the information was "sensitive but unclassified" (Claburn, 2009). The *E-Government Act of 2002* had required that OMB create a public domain directory of public federal government websites, but it took nine years before OMB released its first list and that only listed executive branch domains (Balter, 2011). At the time of the release of that list of 1,759 domains, the White House simultaneously estimated that there were 24,000 individual government websites (Marks, 2011; OMB, 2011). In 2011, the Obama administration set out to reduce the number of government domains by half and, by 2013, it was reported that there were only 993 (Marks, 2013).

Ideally, we would have an up-to-date, comprehensive list of born-digital published Public Information, but we do not. By law, we should have such a list because, in 1968, Congress tasked the Government Printing Office with creating a "comprehensive index of public documents" (44 USC 1710). Almost 30 years after Congress added to that mandate instructions for GPO to "maintain an electronic directory of Federal electronic information" (44 USC 4101), there still is no comprehensive index or directory of federal digital published Public Information. GPO's cataloging and indexing actually decreased steadily during the early days of the web: from 29,000 titles cataloged in 1991 to 19,000 in 2000 (US General Accounting Office, 2001). Between 2005 and 2021, the rate of GPO cataloging remained essentially flat at an average of just over 19,000 titles per year (GPO, Library Services & Content Management, 2015; FDLP, 2016). These numbers are insignificant when compared to reports of hundreds of millions of URLs found on government domains.

Today, however, there is at least some reliable raw data about Public Information on the web. The government now publishes lists of official government domains (Cybersecurity and Infrastructure Security Agency, 2022; GSA, 2022; US Department of Defense, 2022b; US Department of Defense, 2022a). These official sources list the internet domains that the government uses for its official business. And, using a project that began in 2019, the GSA now scans the federal government webspace every week and assembles and publishes a comprehensive list of public federal .gov websites as the "Federal Website Index" (GSA, Digital.gov, 2020; GSA, 2023). As of March 2024, GSA reports that there are between 10,000 and 12,000 federal government websites.

In the remainder of this chapter, we present the GSA data and analyze Public Information on the federal web by using the extensive End of Term (EOT) crawls. These crawls have a specific goal to "archive the US Government web" (End of Term Web Archive, 2024). No other project has such a specific and comprehensive scope of archiving the entire US Government web.

Hosts and Websites

To preserve Public Information published on the web, preservationists need to be able, first, to understand and characterize the federal government webspace. The first task is to identify where on the web the government posts its Public Information.

The vocabulary of the web can be confusing when the same terms are used to mean different things in different contexts. For consistency, we therefore define here how we use four important terms. Our definitions may be slightly different from common usage, but we suggest that they are useful because they can be easily and consistently applied. (For comparison, see the GSA's definitions [GSA, 2021].)

- **Official Domains**: These are the two-level domains (e.g., whitehouse.gov) that the federal government says it uses for official business.
- **Hosts**: This is a subset of Official Domains. These are the two-level domains that host public websites. There are fewer hosts than domains because some official domains are not publicly accessible, and some do not host web content.

- **Websites**: This is a superset of Hosts. It includes all Hosts (because they are websites as well as hosts of other websites) plus all domains of three or more levels on those hosts. Thus, for the National Archives, there is one Host (i.e., archives.gov), but there are 68 Websites including three-level sites (e.g., blogs.archives.gov) and four-level sites (e.g., aotus.blogs.archives.gov).
- **URLs**: A Uniform Resource Locator (URL) is an internet address. A URL can point to the homepage of a website or Public Information hosted by the website.

GSA List of Government Websites

In contrast to the early days of the web, there are now many different lists of government websites. The proliferation of lists and the volatility of websites themselves introduces some subjectivity into producing a definitive, accurate count of federal government websites.

The GSA data provide useful baseline counts. Table 6.1 presents data from the 10,203 websites in its "Unique Final Website" snapshot, which GSA describes as "arguably the best count of federal public .gov websites" (data accessed on March 12, 2024) (GSA, Digital.gov, 2023). We used the data from the GSA to divide the counts of hosts and websites into segments by branch of government and split out from that military websites and hosts that are neither dot-gov nor dot-mil (e.g. fed.us, .com, .org). These will prove useful comparisons to the analysis of web harvesting data that follows.

Segment	Hosts	Websites
Exec (non-military)	683	9,338
judicial	7	9
legislative	50	65
.mil	53	680
.com	90	111
TOTAL	883	10,203

Table 6.1. GSA Unique Final Website snapshot, March 2024.

The End of Term Archive (EOT) Crawls

The EOT crawls generate data about the contents of the federal government web space. Since 2008, a group of organizations have partnered on the EOT crawls in order to capture government web content after each presidential election. Organizations involved in the group planning and harvesting over the years have included: the California Digital Library, George Washington University, the Internet Archive, the Library of Congress, Stanford University Libraries, the University of North Texas Libraries, the US Government Publishing Office (GPO), the National Archives and Records Administration (NARA), and the Environmental Data & Governance Initiative (EDGI).

These crawls have varied in size over the years. The wide variation in number of items these crawls find may reveal at least as much about the different scopes of the crawls as they do insight into the size of the government web. The 2016 crawl, for example, included FTP sites, but the 2020 crawl did not. This probably accounts for fewer "items" in the 2020 data.

EOT year	websites ("items")	.gov URLs
2008	3,305	137,829,050
2012	3,273	109,141,353
2016	53,324	[N/A]
2020	31,103	223,885,260

Table 6.2 Numbers of web sites and URLs harvested by four EOT crawls. Source: http://eotarchive.cdlib.org/background.html

The 2020 EOT data we analyzed was a single file of over 1.2 billion records. This "CDX" (crawler index) file contains one record for each time the harvester visited a URL and includes data about each visit. The data include: URL, MIME-type, file name, size, and a hash digest value (produced by a hash function, this 32-byte string uniquely identifies each digital object downloaded by the crawl) (Internet Archive, 2022).

Each record in the CDX file contains information about one URL, but a URL may have more than one CDX record if it was visited more than once during the crawl, as is typical.

The analysis that follows uses this information to derive some basic facts about the size of the government web and the Public Information available on it. Given these data, we can get some reliable, if preliminary and incomplete, counts that suggest some starting points for preservation planning.

The 2020 EOT was a very broad crawl, capturing many non-federal web pages, non-government web pages to which government websites linked, and social media sites. We used a list of 1,788 federal "hosts" (top-two-level domains such as whitehouse.gov), to create a subset of CDX data only from federal web hosts and their websites. Our federal-only subset contained more than 290 million CDX records.

Although this limited what we analyzed to about 24 percent of the original CDX file, we believe this approach has the benefits of being both comprehensive and precise. It is comprehensive because, according to a 2018 study by the Library of Congress, agencies report that "all" or "nearly all" or "95% or more" of their public information products are disseminated through their websites (US Library of Congress, Federal Research Division, 2018). NARA had similar findings in a 2010 study of agencies use of social media (NARA, 2010). It is precise because it eliminates most of the data that is non-federal. For example, 44 percent of the records for dot-gov URLs in the original CDX file were from non-federal web governments such as states and cities. Although our analysis does not include social media used by government agencies, the same Library of Congress study reported that most agencies reported that they use social media to alert the public to new content on their main websites. From that we conclude that virtually all born-digital federal government Public Information is posted on agency websites and very little unique information is posted elsewhere; at least that was true in 2018. More analysis will be needed to determine how accurate that conclusion is today.

Using data from web harvests to try to characterize the government web is hindered by several problems. First, CDX data can only be used to directly measure the harvest itself, not the web space harvested. The web space harvested may be "deeper" or "wider" than what the crawl harvested.

Second, some websites link to documents only through their internal search features, and those search features can confound harvesting software. Website search features may block harvesters completely or make it difficult for harvesters to find information only available through search forms. Harvesters can get stuck in odd loops that result in both missing some content and re-harvesting the same data over and over again. We found one example of such a loop in the EOT's crawl of the GPO's *Catalog of Government Publications* (CGP) website (catalog.gpo.gov). Almost 22 percent of the EOT 2020 CDX records for the legislative branch (five million out of 21 million records) were from this one site! The CGP is a database of over one million bibliographic records that describe government publications. Those five million CDX records came from less than 30,000 unique files, the rest were duplicates. Almost all of those unique files were web pages of individual records in the CGP database. So, in this case, the crawl over-harvested a few thousand "pages" while missing hundreds of thousands of other, similar pages. This problem was apparently caused by the web harvester interacting badly with the design of the CGP website. This is an extreme, but not atypical, example of one of the key problems web harvesting faces.

Third, harvesters are typically blocked by paywalls. This problem is most notable in the judicial branch where most of its born-digital Public Information is behind a government paywall, PACER (see Chapter 7). The 2020 EOT data that we analyzed did not include that body of information.

Nevertheless, the CDX data provides the best measurement available for determining the size and characteristics of the Public Information on the government web. It provides solid indicators, if not direct measures, of the government web.

More research is needed to explore the government's use of non-official web channels, especially social media platforms and public data-sharing services such as GitHub and Amazon Web Services and Google Cloud.

Nevertheless, we know of no other dataset that is as comprehensive as the EOT CDX data. With these caveats, we do believe that our analysis gives a very good preliminary understanding of the state of born-digital government information as of 2020.

Segments

We created five separate subsets of the CDX data for analysis: one for each of the three branches, a separate subset just for military domains, and a segment for federal government domains outside .gov and .mil (GSA, 2022b).

segment	official domains	hosts	websites
exec	1,142	963	25,221
judicial	24	15	715
legislative	118	83	1,260
.mil	136	130	2,759
.com	368	229	1,148
TOTAL	1,788	1,420	31,103

Table 6.3. 2020 EOT harvest: Numbers of official federal government domains, hosts, and websites, by segment.

Our count found considerably more hosts and websites than GSA found in 2024 (Table 6.1). The difference could be attributed to the counting procedures and definitions used by EOT vs. GSA, or it could be that the government has reduced the number of hosts and websites between 2020 and 2024.

Web Pages and PDFs

The 2018 study by Library of Congress confirms what regular users of government information know: agencies publish all kinds of information in all kinds of organizations, presentations, and formats. The study lists reports, budgets, datasets, court decisions, blogs, social media posts, audiovisual content, and scientific, technical, and regulatory data. Although "MIME-types"—now called "media types" (Internet Assigned Numbers Authority, 2024)—returned by web servers and recorded in the CDX file are not always an accurate method for identifying kinds of web content, they do provide an indicator of format variety. The CDX data for federal domains had nearly 100 different mime-types across the judicial and legislative branches and almost 500 mime-types in the executive branch. Two types, however, were always at the top of the most-used list: HTML and PDF.

We therefore used the MIME-types in each CDX record to count the numbers of "web pages" (HTML files) and PDF files in the EOT 2020 data (Table 6.4 and Graph 6.1). Although the counts of these types of files may not present an equivalent of the number of "documents" or "titles," they do provide a consistent measurement of the scale of content available across very different kinds of websites.

segments	wp + pdfs	everything else
exec	243,865,088	11,839,402
judicial	479,898	94,865
legis	17,718,309	3,572,949
.mil	10,306,073	817,989
.com	1,007,688	314,500
TOTAL	273,377,056	16,639,705

Table 6.4. 2020 EOT. Web pages and PDFs as a percentage of the government web.

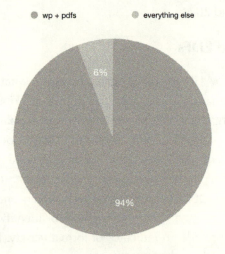

Graph 6.1. 2020 EOT. Web pages and PDFs as a percentage of the government web.

Duplicates

The above counts can be further refined by taking duplicates into account. Web harvests inevitably visit some URLs more than once, and they often download and save the data they find at that address more than once. There are many reasons for this, some intentional and some unintentional. For example, a website may have many web pages that all use the same logo and the same style sheets. In such cases, every time the web-crawling software examines any one of those web pages, it will find links to those same logo image files and CSS stylesheet files and may download and save them. Exactly what gets downloaded and stored depends on how the harvesting software is tuned, how it interacts with the server, and so forth. As an example, we looked at one NOAA logo image file [https://www.noaa.gov/themes/custom/noaa/images/noaa_logo.png]. The Internet Archive and EOT harvesters have encountered this file 3,463 times and downloaded it 343 times. We already eliminated a lot of duplicates of this type by counting HTML pages and not counting the many types of content embedded into them (e.g., image files like this logo file as well as style sheets, JavaScript, etc.). We can go further by using the CDX data to eliminate duplicate web pages and PDFs from our counts.

Another cause for duplicates is redirection, in which many URLs are redirected to a single URL. For example, in 2020 the domain SupremeCourtUS.gov, one of the 24 official judicial domains, apparently did not host any content but simply redirected all requests to SupremeCourt.gov. The EOT-2020 harvest requested 69 different URLs from that unused domain and, each time, was always redirected to the same page (presumably the home page of SupremeCourt.gov). Each of those requests has a record in the CDX file, and each of those records has the same hash digest value, indicating it is identical content regardless of URL originally requested. By examining patterns like this, we can easily find URLs that we counted as web hosts that do not actually host any content.

To determine how much of the EOT 2020 harvest was duplicates, we used the hash values in the CDX records. The hash value is unique for each file harvested. If three CDX records have the same hash value, that represents one unique item and two duplicates of that item. Using this technique, we counted all CDX records and all "duplicate records" and computed the percentage of all records that were duplicates for each segment of the federal government (Table 6.5, Graph 6.2).

segment	total records	unique	duplicate	dup % of total
exec	255,704,490	112,373,291	143,331,199	56
judicial	574,763	482,496	92,267	16
legis	21,291,258	17,855,241	3,436,017	16
.mil	11,124,062	6,934,734	4,189,328	38
.com	1,322,188	1,150,149	172,039	13
TOTAL	290,016,761	138,795,911	151,220,850	52

Table 6.5. 2020 EOT harvest: Numbers of CDX records, unique records, duplicate records, and duplicates as a percentage of all records, by segment.

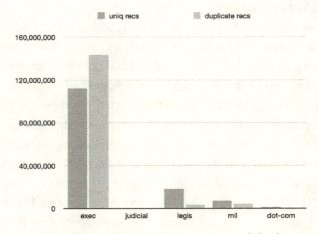

Graph 6.2. 2020 EOT: Percentage of duplicate records

The enormous number of duplicates harvested from the executive branch results in a duplicate rate of over 50 percent for the EOT 2020 harvest as a whole.

This suggests that simply counting all web pages and PDF files as we did above may overestimate the quantity of Public Information available on the government web. We therefore produced counts of unique and duplicate web pages (Table 6.6) and PDF files (Table 6.7) for each segment.

segment	All HTML	Unique HTML	Duplicate HTML	Duplicate % of all
exec	241,831,352	102,546,375	139,284,977	58
judicial	302,337	282,996	19,341	6
legis	16,985,289	16,121,981	863,308	5
.mil	10,159,849	6,502,941	3,656,908	36
.com	939,086	867,723	71,363	8
TOTAL	270,217,913	126,322,016	143,895,897	53

Table 6.6. 2020 EOT. Numbers of all, unique, and duplicate HTML web pages, and the percentage duplicates of all.

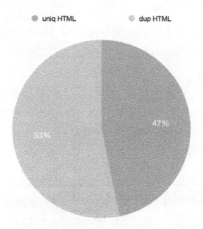

Graph 6.3. 2020 EOT: Unique and Duplicate HTML web pages.

Segment	All PDFs	Unique PDFs	Duplicate PDFs	Duplicate % of all
exec	2,033,736	1,841,723	192,013	9
judicial	177,561	159,532	18,029	10
legis	733,020	682,462	50,558	7
.mil	146,224	134,999	11,225	8
.com	68,602	64,829	3,773	5
totals	3,159,143	2,883,545	275,598	9

Table 6.7. 2020 EOT. Numbers of all, unique, and duplicate PDFs, and the percentage of duplicates of all.

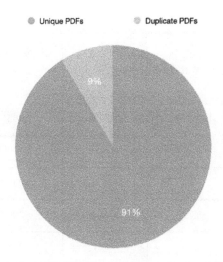

● Unique PDFs   ● Duplicate PDFs

Graph 6.4. 2020 EOT: Unique and duplicate PDFs.

While there was only a relatively small amount of duplication in most of the EOT 2020 harvest, the enormous number of duplicate web pages being harvested from the executive branch skews our earlier counts significantly. This suggests that we can get a more accurate estimate of the size of the government web by using the counts of unique PDF and HTML files.

When we factor duplicates into our earlier analysis of the dominance of web pages and PDF files on the government web (Table 6.4 and Graph 6.1), we find the pattern holds.

Measuring unique HTML and PDF files as a percentage of all unique CDX records (Table 6.8 and Graph 6.5) reveals that, when duplicates are removed from the equation, PDF and HTML files still comprise the great bulk of the federal government web.

Segment	unique records	unique html + pdf	everything else	% html + pdf of all records
exec	112,373,291	104,388,098	7,985,193	93
judicial	482,496	442,528	39,968	92
legis	17,855,241	16,804,443	1,050,798	94
mil	6,934,734	6,637,940	296,794	96
dot-com	1,150,149	932,552	217,597	81
totals	138,795,911	129,205,561	9,590,350	93

Table 6.8. 2020 EOT. Unique PDF and HTML files as a percentage of all unique records.

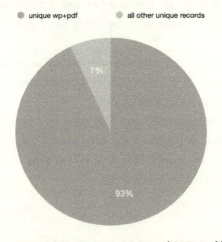

● unique wp+pdf   ● all other unique records

7%

93%

Graph 6.5. 2020 EOT. Unique PDF and HTML files as a percentage of all unique records.

More than 90 percent of the government web is PDF files and web pages. Much of the remaining 7 percent or so comprise various kinds of content, much of it being elements embedded in web pages (images, JavaScript, CSS). That 7 percent of the government web comprises some hundreds of other different formats.

We can also use our counts of unique and duplicate records to get a better understanding of the number of websites that host most of the Public Information. We used counts of unique records (i.e., unique content of any file-type) to identify websites by size using cutoffs of 100 and 1,000 records.

- **Large**. Websites with more than 1,000 unique records.
- **Medium**. Websites with between 100 and 1,000 unique records.
- **Small**. Websites with fewer than 100 records.

segment	Large websites	Medium websites	Small websites
exec	1,307	2,861	20,347
judicial	132	201	360
legis	168	514	451
.mil	184	820	1,661
.com	91	158	846
TOTAL	1,882	4,554	23,665

Table 6.9. 2020 EOT harvest: Numbers of websites, by size and segment.

This table suggests that most Public Information resides on those 1,882 large websites. To check to see if that is correct, we added up all the unique web pages and unique PDF files on large, medium, and small websites.

content type	1,882 large	4,544 medium	23,666 small
web pages	125,640,911	610,274	70,812
pdf files	2,758,435	121,108	3,981
totals	128,399,346	731,382	74,793
percent	99.38%	0.57%	0.06%

Table 6.10. 2020 EOT harvest: Numbers of web pages and PDF files, by website size.

This simple count of the major content types suggests that more than 99 percent of Public Information is posted on fewer than 2,000 websites.

Conclusions

We can make some preliminary conclusions from this analysis of the government web spaces.

- **Hosts not Websites**. The government web is large but far from infinite. Many of the very large counts of "websites" (whether using our definition or others) probably overestimate the size of the government web. Almost all the content we found was on just 1,882 "large websites," most of which are "Hosts" (which are also websites). "Websites" come and go, but Hosts are much more stable. We examined many estimates of the size of the government web over the last 20 years and found that all of them included a figure around 1,700 to 1,800 domains. This seems to be a pretty stable figure.

- **Text not A/V**. Although there is a significant amount of born-digital content being produced by the government in non-text formats (e.g., audio files, video files, spreadsheets, numeric data), the vast majority of content is text-based.

- **Duplicates**. The presence of duplicates in web archives can inflate estimates of the size of the government web and the amount preserved. Focusing on unique content rather than on the number of files or bytes gives a better indication of something closer to "titles" or "volumes" that need to be preserved. Reports by archives of the number of bytes or files they have stored, without accounting for duplication (or replication or backup) is too imprecise to provide a realistic measure of the success of their preservation activities.
- **The Executive Branch is the Leviathan**. The executive branch is by far the biggest born-digital publisher in the government. This should not be surprising since it is also the largest branch of the government, but the scale is worth noting. By our measurement, almost 90 percent of government PDFs and 65 percent of government web pages were published by the executive branch.
- **Selection**. By focusing in this analysis on the bulk of digital government information (PDFs and web pages), we do not imply that the remainder (e.g., non-textual content, content on non-official government websites including social media sites, content on small websites, content stored in commercial "cloud" services, etc.) is unimportant or insignificant. Neither do we mean to imply that only web pages and PDF files need to be preserved. This analysis is only meant to give a preliminary overview of the scope of born-digital content. This analysis does suggest something important, however: focusing selection on particular segments of Public Information might be both more efficient and more effective than less-focused, broad web harvests. Focused selection may be more manageable in terms of scale, more efficient in terms of preservation resources, and more effective in terms of providing access to the content needed by specific Designated Communities. Indeed, some of the smaller websites and smaller content types may be excellent targets for relatively small-scale content acquisition projects.

CHAPTER 7
GPO's GOVINFO

In this chapter, we offer some preliminary analysis of what is being preserved and what is not being preserved by examining the contents of GPO's preservation repository, GOVINFO. We chose GOVINFO because it is the only repository preserving government information that has been certified as an ISO 16363 Trusted Digital Repository that complies with OAIS (PTAB Primary Trustworthy Digital Repository Authorisation Body Ltd, 2018). Our focus in this analysis is on born-digital content, *not* paper publications that have been digitized.

We had only indirect access to the contents of GOVINFO, so our analysis and the conclusions we draw cannot be complete or precise, but we believe they can be accurate enough to identify existing gaps in preservation. Ideally, we hope this preliminary work will inspire producers and preservationists to envision, design, develop, and provide the infrastructure needed to more accurately and comprehensively understand preservation gaps and identify targets for preservation. In Chapter 21, we introduce the idea of a Metadata Repository, which can, among other uses, provide a source for such measurement data.

Preservation repositories could provide consistent metrics for their holdings, but they do not. In fact, it is difficult for them to do so because there is no universal metric for counting or even identifying digital content. Paper-era definitions tended to equate content with its packaging (i.e., a book title with its physical volume), which led to familiar metrics such as volume count. Finer-grained content identification such as article titles in issues of journals were also discrete and relatively unambiguous. In the digital age, content and its packaging are not inherently bound together in the same way and content delivery is (at least potentially) much more fine-grained than in the paper era. Born-digital content also presents the opportunity to divide and combine content in unique ways for each user or each user interaction. One user, for example, might want an issue of the *Federal Register*, but most users will want just a single announcement, notice, or regulation from inside an issue. Other users may want the series of Notices and Announcements about a single proposed regulation that were made and published over a period of months in many different issues of the *Federal Register*. In the digital age, bundling and delivering content in these different ways is possible by decoupling the way information content is produced, transmitted, stored, and delivered.

Digital technologies also give us new digital objects to manage. In terms of preservation metrics, we are left with a sometimes-bewildering array of kinds of things to identify and count: titles, files, web pages, websites, bytes, database records, and various kinds of packages such as OAIS's SIPs and AIPs and DIPs (see Chapter 15) and web-harvesting software's WARCs (US Library of Congress, 2020a).

Without consistent and well-reported metrics, preservation planners are left with using whatever information is publicly available to make general characterizations about production and preservation of digital content and get a preliminary understanding of preservation gaps. Because of the limitations of available data, we spell out our methods in some detail in this chapter in order to make the conclusions we draw (and their limitations) as clear as possible. We also hope that, by explaining our methods, we will inspire producers and preservationists to develop better reporting mechanisms to improve the accuracy and comparability of future analyses.

Contents of GOVINFO

Regular users of government information will be familiar with GPO's govinfo.gov website. That website provides access to the contents preserved in GPO's digital repository, which is also known by the name GOVINFO. The website also provides links to content that is not in the GOVINFO repository. The "Browse by A-Z" page of the govinfo.gov website, for example, points to content held and managed by other institutions. (Although some of these have an explicit preservation "partnership" agreement with GPO – e.g., reports from the Government Accountability Office [GAO], we do not analyze content held by those partners here because it appears to us that few if any of those partners are doing ongoing, active preservation of new born-digital content, which is the focus of this chapter.) Thus, we limit our analysis here to the contents actually held and managed by GPO in the GOVINFO repository. To avoid confusion, we refer here to the repository itself as GOVINFO and the public website that provides access to the repository as govinfo.gov.

As noted above, different repositories use different terms to describe their content and report on their size. Some terms are unambiguous—like "terabyte," which denotes 1,000 gigabytes—and are familiar even to non-preservationists at least in a general way. But byte-count is not a very useful metric for understanding the contents of a repository without a clear understanding of what is being counted (e.g., all content including replication and backups? unique content?). Other terms like "file" and "folder" are familiar to some casual computer users, but many such terms are actually vague metaphors rather than precise engineering terms (Chin, 2021). Other terms are new and require careful definition.

The production of government information is still, in many ways, tied to the print-era concepts of books and issues of serials, and that gives us metrics tied to those concepts. The *Federal Register*, for example, is still produced as a daily serial publication, with each issue being assigned a volume number, an issue number, and page numbers. But the shift to digital publishing provides new ways to store, index, and deliver content, resulting in new terminology to describe these new containers of digital objects.

To gather metrics on the content of the GOVINFO repository, we used govinfo.gov's Applications Programming Interface (API), which is available to registered developers, and its search and browse tools, which are available to the general public. These tools provide metrics using the terms Packages, Granules, and Records. A "Package" is roughly equivalent to one bound printed volume; "Granules" are roughly equivalent to parts of volumes; and "Records" are roughly analogous to library catalog records, which sometimes point to packages and sometimes to granules (GPO, 2021e; GPO, 2018c; GPO, 2021c). The metrics we gathered in the tables below varied a bit over the time we gathered them. The analysis was done in early 2022.

Packages (Volumes)

GPO uses the number of "Packages" as a metric in its promotional materials and annual reports, noting that GOVINFO has more than two million packages and adds 100,000 to 200,000 packages per year (GPO, 2021d; GPO, 2021b). The repository contains both born-digital content and older content that was issued in print and has been digitized—mostly into PDF files—and these counts of Packages include both kinds of content. Analyzing the content of GOVINFO by century of publication shows that most of the packages (93 percent) were published in the 21st century and are, presumably, born-digital content. (These figures come from using the website's "Browse by Date" option, which returns results as "Packages" [GPO, 2022a].) Presumably, much, but far from all, of the 7 percent of the repository that was published in the 20th century was also born-digital.

date published	number packages (vols)	percent of repository
18th century	2	0.0001%
19th century	9,456	0.5%
20th century	134,527	7%
21st century	1,894,616	93%
TOTAL	2,038,601	100%

Table 7.1 Number and percent of Packages (volumes) in GOVINFO, by date published.

Although characterizing those two million Packages as "volumes" provides a familiar image of the size of GOVINFO and suggests that we can compare that size with the size of traditional libraries, that picture is misleading. GOVINFO Packages vary greatly in size: from court opinions of fewer than 10 pages to the *Senate Manual* of more than 1,000 pages. Breaking down the counts of Packages in the repository by branch of government reveals that almost 75 percent of all those Packages are small-sized court opinions. That means that only about 500,000 of the two million Packages would have been published in traditional "volumes" in the print era. This count also reveals that a very small amount (2 percent) of those "volumes" are from the executive branch.

branch	packages	percent of repository
exec	51,577	2%
judicial	1,605,850	74%
legis	518,543	24%
TOTAL	2,175,970	100%

Table 7.2. Number and percent of Packages in GOVINFO by branch.

Granules (Parts of Volumes)

GOVINFO breaks up many publications into "granules," which are the discrete content that users often seek. For example, few users want an entire issue of the *Federal Register*; they want to find, retrieve, and read a particular Notice or Rule, and each of those is one "granule" in GOVINFO. For example, the January 3, 2022, issue of the *Federal Register* is stored as 89 PDFs: one PDF of the entire issue and 88 small PDF files, each for a specific entry in that issue. The small PDFs contain parts of the issue such as the Agriculture Department's "Notice of Proposed New Fee Sites" (one page) and the Coast Guard's "Port Access Route Study: Northern New York Bight" (three pages). Even some large documents, such as the *Senate Manual*, are broken into granules. For example, in addition to being available as a single, 1,000-page PDF file, the 110th Congress *Senate Manual* is also stored as 179 smaller PDF files ("granules") such as "Rule I: Appointment Of A Senator To The Chair" (two pages) and "Rule XXX: Executive Session—Proceedings On Treaties" (two pages).

Viewing the number of granules per branch gives a different perspective on the content of GOVINFO. (In the table below, we used the API to obtain counts of granules and aggregated those counts into totals for each branch of government.)

branch	granules	percent
exec	7,464,776	50%
judicial	3,465,443	23%
legis	3,927,386	26%
TOTAL	14,857,605	100%

Table 7.3. Number and percent of granules in GOVINFO, by branch.

Unfortunately, not all large packages are broken down into smaller granules. For example, the *Journal of the Senate* and the *Statutes Compilations* and the *Congressional Serial Set* have no granules. So, while a count of granules reveals more detail about the content of some titles, it masks details of others. For example, the *Code of Federal Regulations* has about 5,000 "volumes," but more than six million "granules." Compare that to the Serial Set, which has 14,000 volumes, but no granules at all. The fact that so many of the executive publications are broken into many granules, while small judicial opinions are not, accounts for the big differences in percentage for the judicial branch between Tables 7.2 and 7.3.

Records (Metadata)

A third way of characterizing the content of GOVINFO is to count "Records." These are metadata (descriptions of content) that point to content. Searches of the public website govinfo.gov return results as Records, most of which point to granules but some to packages.

branch	records	percent
exec	1,244,736	19%
judicial	3,465,469	52%
legis	1,940,273	29%
TOTAL	6,650,478	100%

Table 7.4. Number and percent of records in GOVINFO, by branch.

Growth of GOVINFO

GPO's 2020 Annual Report says that it added 212,000 packages (volumes) to GOVINFO in FY 2020. By using the API, we were able to gather data for the number of new packages added between early January and late May 2022. During this roughly four-month period, 52,325 packages were added to GOVINFO, which implies an annual rate of expansion of roughly 200,000—a count that is similar to the additions in FY 2020.

The API provides a package count for each of 38 "Collections," which are sets of content that have a consistent format. Typically, each collection is a publication (e.g., the *Senate Manual*) or serial (e.g. the *Federal Register*) or series such as "Congressional Hearings" (GPO, 2021e). Some collections seem to be static in size; that is, no new content is being added to them. These appear to be historic collections that were imported in bulk, perhaps in some cases from digitizing print publications, and no new content has been added to them for some time. For example, the *Education Reports from ERIC* (1995-2004) and *GAO Reports and Comptroller General Decisions* (1994-2008). Additions are made to some collections infrequently because of the nature of the publication. An example is the *Economic Report of the President*, which is published annually.

The "United States Courts Opinions" collection is, by far, the fastest growing of the 38 collections in GOVINFO, comprising 80 percent of the packages added during the four-month period we measured. Most of the rest of the growth of GOVINFO during this period was confined to just five other collections.

GOVINFO collection	packages added January to May 2022	percent
United States Courts Opinions	41,906	80%
Congressional Bills	3,406	7%
Congressional Bill Status	2,861	5%
Congressional Serial Set	1,070	2%
Congressional Reports	871	2%
Congressional Hearings	770	1%
Compilation of Presidential Documents	395	0.01

Table 7.5. Count and percent of packages of fastest growing GOVINFO collections, January-May 2022.

Missing Content

While all of these comparisons describe what is being preserved, none reveals what is not being preserved. Since there is no comprehensive catalog or listing of born-digital government information, we cannot provide an exact metric of unpreserved content. An analysis of the metrics we do have, however, suggests that a very significant amount of born-digital content is not being preserved in GOVINFO.

GOVINFO delivers almost all content to users as PDF files. The repository contains a little over two million "Packages," each of which contains one or more PDF files that roughly comprise a single "volume."

Although GOVINFO contains some PDFs from the 18th through 20th centuries, the great bulk of the content (93 percent) is from the 21st century. It has taken GOVINFO 22 years to accumulate about two million PDFs from the 21st century. The EOT crawl found almost three million PDFs still on the web in 2020. How many of those three million PDFs are new and how many have been on the web for years? We do not know. We also do not know how many PDFs have been posted to the web and then withdrawn or changed over the years. But one study determined that 83 percent of the PDF files present in the 2008 EOT crawl were missing in the 2012 EOT crawl (Phillips, 2016). One explanation for an 83 percent loss of content is that a great deal of content posted on the web is withdrawn either during a four-year administration or by a new administration. This inference requires further study.

To extend this general analysis, we analyzed the contents of GOVINFO from the perspective of each of the three branches of the federal government and report the results in the remainder of this chapter. To gain a better perspective on the effects of the digital shift, we relate the histories of the changes brought to each branch by the shift to digital publishing.

Legislative Branch

Legislative branch publishing has been fairly stable over time, including after the shift to digital publishing. Most of the traditional publications of Congress are still handled by GPO and go into GOVINFO for preservation. Web publishing has, however, added uncertainty to preservation of legislative branch Public Information.

The legislative branch includes Congress and seven agencies (the Architect of the Capitol, the Congressional Budget Office, the Government Accountability Office, the Government Publishing Office, the Library of Congress, the Congressional Research Service, and the

United States Botanic Garden). GPO publishes official publications of Congress and must do so by law (44 USC 7). This publishing continued after digital publishing was introduced. As a result, some important legislative branch publications have a direct path into GOVINFO for preservation.

Indeed, 24 of the 38 "Collections" in GOVINFO are publications of the legislative branch and almost all are publications of Congress itself. These include nine series (Congressional Bills, Bill Status, Calendars, Documents, Hearings, Committee Prints, Reports, Public and Private Laws, and GAO Reports), and more than a dozen titles, including: Statutes Compilations, the *Congressional Record* (daily, bound, and index), the *Congressional Directory*, *Economic Indicators*, the *Journal of the House of Representatives*, the *House Rules* and *Manual*, the *History of Bills*, the Serial Set, the *Journal of the Senate*, the *Senate Manual*, the *Statutes at Large*, and the *United States Code*. Altogether, these account for half a million Packages or about 22 percent of the content of GOVINFO.

In terms of growth, Legislative branch Collections were 7 of the top 10 fastest growing Collections in GOVINFO. Over the period we analyzed (January–May 2022), legislative packages also accounted for almost 20 percent of the content added.

This record of stability and successful digital preservation must be tempered by the fact that the legislative branch also publishes other content directly to the web. In our analysis of the 2020 EOT crawl (Chapter 6), we found 83 legislative host domains and more than 1,000 websites (168 of them "large" sites) containing more than 16 million unique web pages and more than 680,000 unique PDF files.

At least some, if not most, of this web content may be early versions or duplicates of content that gets officially published later. Examples of that would include testimony at committee hearings, the original text of bills, and Acts passed.

Some Congressional web content is explicitly rejected by GPO as "out of scope" of its Title 44 remit (GPO, 2020c). GPO's strict interpretation of the law prevents it from accepting web-published content created by an individual Member of Congress if that publication was not conducted "under the auspices of a congressional committee." An example of such an excluded publication is the *Social Media Review* study conducted by Zoe Lofgren (Representative of the 19th Congressional District of California), which contains lists of public social media posts from Members of the US House of Representatives who were sworn in to office in January 2021 and who voted to overturn the 2020 presidential election (Lofgren, 2021).

GPO does have a small web-harvesting program, the "FDLP Web Archive" (FDLP-WA). In our analysis of FDLP-WA content in early 2022, we found that only 14 of the 254 collections were legislative branch websites. We examine the FDLP-WA program in detail in Chapter 9.

Executive Branch

As noted above, we found that only 2 percent of the packages in GOVINFO were from the executive branch. That suggests that a large quantity of executive branch Public Information is missing from GPO's digital repository. There are two reasons to suspect that this is the case.

The first is that it seems unlikely that the largest branch of government would produce the smallest quantity of content. There are 195 top-level offices in the executive branch employing over two million people, compared to 37 offices and 60,000 employees in the other two, combined (US Office of the Federal Register, 2020; Congressional Research Service, 2021). Because there are so many more offices and people creating information in the executive branch than in the other two, one would expect it to dominate the content in GOVINFO if all branches were equally well represented there.

The second reason to suspect that large quantities of executive branch content is missing from GOVINFO is the executive branch has a long history of refusing to provide its content to GPO (Jacobs, 2017). As we mentioned in Chapter 5, the 1983 Supreme Court *Chadha* decision gave the executive branch legal permission to withhold its publications from GPO and the FDLP (Congressional Research Service, 2017). Even before *Chadha*, some estimates of the Public Information not making it to GPO were as high as 85 percent (Federal Documents Task Force of the ALA Government Documents Round Table, 1973). After *Chadha*, official estimates were as high as 50-78 percent (Baldwin, 2003). It is reasonable to assume that the situation has worsened in the 21st century since the 2018 Library of Congress survey found that most agencies were publishing 100 percent of their Public Information digitally (2018).

The data from our analysis of the 2020 EOT data (Chapter 6) appears to verify this. More than 80 percent of the (freely available) born-digital content published by the government is published by the executive branch (Table 7.6).

segment	total unique web pages and PDFs	percent
exec	104,388,063	81%
judicial	442,527	0.3%
legis	16,804,442	13.0%
military	6,637,938	5%
.com	932,551	1%
TOTAL	129,205,521	100%

Table 7.6. 2020 EOT: Number and percentages of total unique web pages and PDFs, by segment

Executive Branch Publications in GOVINFO

We analyzed the executive branch content in GOVINFO and discovered that more than half of executive branch packages and more than 99 percent of the executive branch granules are contained in just 10 publication titles. The reason for this is that most of the executive branch content in GOVINFO is in the few titles that GPO still publishes. This matches the pattern we found for legislative branch content. The executive branch titles include ongoing serials such as the *Federal Register* and the *Compilation of Presidential Documents*, the Budget, the *Economic Report of the President*, and *Public Papers of the President*. These are significant collections of important titles, but they are not everything the executive branch publishes.

The fact that there are so few executive branch titles in GOVINFO suggests that the vast bulk of executive branch Public Information that is digitally published directly to the web is simply not in GOVINFO. Indeed, GOVINFO's one million executive branch "records" is only about 1 percent of the number of executive branch web pages and PDFs we counted in the 2020 government web space in Chapter 6.

To further analyze this, we used the advanced search capabilities of the govinfo.gov website to examine executive branch publications in GOVINFO that were published during the first 22 years of the 21st century.

GOVINFO contains about one million Records for executive branch publications of the 21st century, but 99 percent of those are "granules" from just five publications. Of the remaining 8.000 records (less than 1 percent of all executive branch publications in GOVINFO), almost all come from just three agencies: the National Institute of Standards and Technology (NIST), the Railroad Retirement Board, and the Education Resources Information Center (ERIC), and two of those seem to be static Collections with no current publications being added (NIST, none after 2019; and ERIC, only four added after 2003).

source	Records
5 publications (*Federal Register, Code of Federal Regulations, Compilation of Presidential Documents, Public Papers of the Presidents, List of CFR Sections Affected*)	965,654
ERIC	548
NIST	3,069
Railroad Retirement Board	1,973
all other agencies	2,294
TOTAL	976,834

Table 7.7. Number of 21st century executive branch records in GOVINFO, by source.

Test Case: The Treasury Department

As a second analysis of executive branch publications, we compared the content in GOVINFO with the actual born-digital publishing output of a cabinet-level executive branch agency. We chose the Treasury Department because, when we did this analysis, it was the only cabinet-level agency that the govinfo.gov website listed on its page for browsing "Executive Agency Publications." We compared the contents of GOVINFO with the content in the 2020 EOT Crawl (Table 7.8).

Limiting our search to publications from the 22-year period 2000–2021, the govinfo.gov website listed only 17 publications in the GOVINFO archive from the Treasury Department, none more recent than 2008. Using data from the 2020 EOT crawl, we found over 100,000 unique web pages and over 25,000 unique PDF files on Treasury Department websites.

type	GOVINFO	EOT-2020
web pages	0	166,567
PDFs	17	25,454

Table 7.8. Number of Treasury Department Publications in GOVINFO, 2000–2021, and EOT-2020.

Summary

This analysis is far from a complete examination of the executive branch contents of GOVINFO and, by the time you read this, the contents will have changed. We do believe, however, that it is an accurate characterization as of early 2022, and we have not yet seen anything suggesting that this pattern will change soon. We believe this analysis accurately demonstrates two things. First, the born-digital executive branch content that is making its way into GOVINFO is from only a very few publications. Second, GOVINFO is getting almost none of the content that executive branch agencies are publishing directly to the web.

Judicial Branch

The Public Information of the judicial branch has a complex structure and history that may seem confusing and even paradoxical to the newcomer. The shift to digital publishing has, of course, affected both the access to and the preservation of this body of information. An understanding of the history of changes in access and preservation brought on by the digital shift provides a context for preservation in the future.

The Paper Era

We might summarize the print era of judicial Public Information as being largely privatized by the big legal publishers (Truesdell, 2010; Chandler, 2000; Biscardi, 1993). But the full story is more complex and leads to vast changes in the digital age.

In the print era, access to the published opinions of federal courts was largely limited to patrons of those libraries that could afford to subscribe to expensive commercial publications of the opinions. Preservation was accomplished by the print-era infrastructure of many copies of bound volumes being distributed to and saved by many libraries. This oversimplifies a complex infrastructure but highlights two essential features of the era. First, distribution of judicial Public Information was handled almost exclusively by the private sector. Second, although distribution was commercialized, libraries preserved the material by purchasing, acquiring, and retaining those publications for long-term access for their communities. But there is much more to the story.

In the print era, both access and preservation were divided into three different models based on three different kinds of Public Information: traditional government publications, "published" court opinions, and case materials including "unpublished" opinions

Traditional Government Publications

Some judicial branch Public Information (perhaps less than 1,000 titles) was part of the GPO/FDLP infrastructure that provided free public access and, by default, long-term preservation.

The bulk of judicial Public Information that was traditionally published through GPO came from the administrative offices of the judicial branch: the Judicial Conference, the Administrative Office, the Federal Judicial Center, and the US Sentencing Commission. These offices are responsible for setting guidelines, recommending and making policy, administering policies and guidelines, and providing educational materials for judges, court personnel, and the public (Administrative Office of the US Courts, 2022a). They produce manuals, journals, statistical reports, and other publications about the functioning of the judicial branch. Some of these were published by GPO and distributed to Federal Depository Libraries (e.g., the *Reports of the proceedings of the Judicial Conference of the United States*) (Judicial Conference of the United States, 1963).

A few individual courts also did a small amount of traditional publishing through GPO. Their publications included informational papers, annual reports, and court rules. Examples of these include *Federal rules of bankruptcy procedure and local bankruptcy rules and forms* (US Bankruptcy Court [Colorado], 1995) and *Jurisdiction of United States Court of Claims* (US Court of Claims, 1970). Some of these traditionally published titles were distributed by GPO to Federal Depository Libraries and preserved there.

The judicial branch is about the same size as the legislative branch with about the same number of employees (Congressional Research Service, 2021) and actually more "top level Named Agencies" (US Office of the Federal Register, 2020), but its traditional publishing output has always been tiny in comparison. Altogether, by our count, only about one-tenth of 1 percent of records in the *Catalog of Government Publications* are from the judicial branch (GPO, 1995a). But the judicial branch also produces large quantities of Public Information that were either commercially published or never traditionally published at all.

Published Court Opinions

The courts themselves (13 appellate courts, 94 district courts, 94 bankruptcy courts, the Court of International Trade, the Court of Federal Claims, and The Supreme Court) issue opinions. Opinions that the judge considers to be significant (or "precedential") are published in "case reporters"—bound volumes with the opinions printed in chronological order. Most precedential opinions are produced by appellate courts and are published by commercial publishers (Albaum, 2002). Only three of the 204 judicial branch federal courts have been regularly published by GPO and distributed to depository libraries: the Supreme Court, and a small number of opinions from the trial courts of customs and claims (US Court of International Trade, 1984; US Court of Claims, 1982) (US Supreme Court, 2025).

Commercially published court opinions gained short-term and long-term access and preservation through purchase by law libraries (Zhu, 2012). Unfortunately, subscriptions to those publications are expensive. In 2022, a subscription to the Thompson West *Federal Reporter* Third Series (which covers opinions by the 13 circuits of the United States Court of Appeals) was more than $4,000 per month (Thompson Reuters, 2022). Preservation was dependent on law libraries that could afford to purchase Public Information from the commercial publishers. Because of the value of a complete record of court opinions to a stable commercial market of law offices and law libraries, the commercialization of legal information effectively preserved this significant, important body of public judicial information. Unfortunately, this also meant that it was not easily and broadly available for free public access (Zhu, 2012). FLDP libraries are required by law to make content deposited through the depository system available to the general public. Libraries that purchase commercial publications—including commercially published case reporters—are not required to make those publications available to the general public. Some public law libraries do make their collections open to the public, however.

Case Files and Unpublished Opinions

The vast majority of Public Information created by the judicial branch are the courts' case files and unpublished opinions.

Case files are all the documents filed with the clerk of the court, plus the transcripts of judicial proceedings and "docket-sheets" that catalog the files for each case. Case files can include exhibits, briefs, motions, orders, and judgments, as well as the final opinion of the court (Federal Judicial Center, Federal Judicial History Office, 2010; Administrative Office of the US Courts, 2016). These records of court proceedings and unpublished opinions are implicitly understood to be Public Information (Hernon & Relyea, 1995; Martin, 2008) and the courts made the right of access to judicial information explicit in the 1970s and 1980s (Schultze, 2018).

Most court opinions are classified by the courts as not precedential and therefore remain as unpublished parts of case files. There is no definitive accounting of how many opinions are left unpublished, but estimates range from 75 percent to 97 percent of all opinions (Griffith, 2015; Gerken, 2004; Keele, 2012). These unpublished opinions along with other case materials were generally left as paper Records, stored in the courthouses where the cases were heard. As "Public Information" they were usually available for examination at the courthouses, but they were, in general, not distributed (Administrative Office of the US Courts, 2000). For these public Records, the courts create Records Schedules that determine their ultimate disposition (retention or discarding or transfer to NARA for permanent retention) (Federal Judicial Center, Federal Judicial History Office, 2010).

The cost to the judiciary to store paper Records before their final disposition runs to millions of dollars per year, and that, inevitably, affects what gets preserved. In 2013, the Administrative Office of the US Courts announced that it had changed its Records Schedules to allow it to discard millions of records and save the Judiciary more than $1,000,000 per year (Administrative Office of the US Courts, 2013b; Schultze, 2018). As noted earlier, only 1 to 3 percent of federal Records are scheduled for permanent preservation (NARA, 2020b).

In the paper era, the bulk of the courts' Public Information remained unpublished, undistributed, unpreserved, and, though technically available, practically inaccessible.

The Digital Age

If the paper era was characterized by the commercialization of the published opinions of federal courts and the relegation of most other court-produced Public Information to the status of paper Records with limited preservation and limited access, then the digital age has certainly changed things—a lot. So much information that used to be difficult to access (case records, unpublished opinions) is now accessible online. And much more Public Information is being preserved for long-term, free public access. But online availability does not mean access is either easy or free. In the digital age, there is still published and unpublished information, but the boundaries between them are now blurring, and access to this information and preservation of it has improved. The history of the digital shift demonstrates how laws and bureaucratic decisions can affect access and preservation—positively and negatively.

Commercial publishers adapted to the digital age early on by creating online databases with sophisticated features for searching their traditional body of published opinions. Lexis and Westlaw were introduced in the mid-1970s and expanded greatly by the mid-1980s (Harrington, 1984).

The changes to official public access began in the late 1980s when the Judicial Conference of the United States, which oversees the administration of federal courts, authorized a pilot program of "electronic access" for a dozen of the 204 federal courts. Soon after that, Congress authorized public access to information "through automatic data processing equipment" but directed the courts to create a system that could support itself with fees. This resulted in courts creating fee-based, dial-in bulletin board services (GPO, 1996b, pp. A-123; Jacobs, 1990; Wikipedia, 2025). By the mid-1990s, 180 of the federal courts offered some sort of fee-based public access to some court documents (Moreland, 2021). Making this system self-supporting with fees for public access has proved to be as controversial with the public as it has been lucrative for the courts (Administrative Office of the US Courts, 2000; Hughes, 2019).

That pilot project ushered in public access to digital court documents in a small way, but the big change started in 1995 when the Administrative Office of the US Courts, which provides technology and management support to federal courts, introduced an early prototype of a system for managing court documents digitally: the Case Management/ Electronic Case Files (CM/ECF) system. This was not a public system nor was it designed for public use. It was an administrative system for use by the courts, judges, court staff, and litigants. It was, essentially, simply a digital version of the old, paper-based court administrative procedures (Schultze, 2018). Each court ran its own copy of the CM/ECF software, which allowed case documents to be filed (ECF) and managed (CM) electronically. The goal was to improve the efficiency of the administration of court business, but it had a side effect on public access to court records. Paper Records that had been accessible only by visiting local courthouses were becoming digital Records that were potentially accessible from anywhere. Adoption of the CM/ECF system was phased-in over a period of more than 10 years. By 2005, 80 percent of the federal courts were using the software, and by 2011, all were (Greenwood & Bockweg, 2012). As its use expanded, so did the case materials included in the system—from simple dockets to "images" of case documents (Martin, 2018).

At the same time that electronic case management was being adopted by more courts, the courts' Electronic Public Access (EPA) program was evolving from primitive bulletin board systems to "voice response systems" and, by 1998, to the web (Administrative Office of the US Courts, 2000). By then the system had become known as the "Public Access to Court Electronic Records" system or "PACER." This was a single public interface to the distributed CM/ECF systems. PACER is not a centrally managed database but a gateway to more than 200 separate court databases. The individual courts still run their own CM/ECF software locally, maintaining both operational control of their local systems (even down to which version of the CM/ECF software they run) and managerial control over data entry and data management. PACER allows registered users to search for cases in one or more courts and provides a nationwide "case locator" for finding which court has a case (Administrative Office of the US Courts, 2022b). Although it makes case materials much more accessible, PACER also has barriers that inhibit access. It has a very complex system of user fees, which includes fees for searching as well as for retrieving documents, and it requires users to register even to search the system (Administrative Office of the US Courts, 2025; 2022c). Its search function has been described as "time-consuming and expensive," "awful," "terrible," "tortuous," and "difficult to navigate but also costly and tedious to even enter" (Keele, 2012; Schwartz, 2009; Raymond, 2021; Williams, 2020).

Although these records are Public Information and available on the web, they are behind a paywall. This means they are largely out of the reach of web-harvesters.

The shift to electronic case files affected how the courts treated the information. For years, courts had referred to "published" and "unpublished" opinions and treated them differently legally. Some circuit courts did not permit "unpublished opinions" to be cited in litigation. But as digital case files made unpublished opinions almost as easy to find and access as published opinions, the term "unpublished" became misleading (Stroud, 2015). In 2006, the Advisory Committee on Appellate Rules of the Judicial Conference of the United States approved a new rule of the *Federal Rules of Appellate Procedure*, which required appellate courts to permit "unpublished" opinions to be cited (Rule 32.1) (Federal Judicial Center, Federal Judicial History Office, 2010; Gerken, 2004).

Even as the courts were changing the way they treated case files, Congress changed the law affecting access to those records. With the *E-Government Act of 2002*, Congress introduced a new category of judicial Public Information: "written opinions." The *Act* requires federal courts to provide web access to "the substance of all written opinions issued by the court, regardless of whether such opinions are to be published in the official court reporter, in a text searchable format" (Section 205(a)(5)). PACER essentially already provided much of the functionality that the Act required. (The courts interpreted the law to mean that the opinions delivered must be text-searchable after retrieval, not that the system must provide a full-text search of all opinions—which it does not.)

Although one might think that all opinions that are written down would legally fit into this category of "written opinions" described by the *E-Government Act*, that is not true in practice. For an opinion to be treated legally as "written," it must be tagged as written in the court's CM/ECF electronic case management software. Since each court manages its own CM/ECF data using its own court rules and bureaucratic management procedures and following the instructions of each individual judge for each case, this means some opinions are digitally available in CM/ECF systems and accessible through PACER but are not tagged in that system as "written opinions."

This technical/bureaucratic implementation of the law has practical implications. In 2005, the courts made "written opinions" available to the public without a fee and provided a free search feature for finding opinions (Administrative Office of the US Courts, 2005). But, since not all opinions (including "published" opinions) are necessarily tagged as "written" (Administrative Office of the US Courts, 2005; Minick, 2011), some opinions fall through the cracks and are still not freely available.

Although the Act did not require free access to court opinions, it did change the wording of an earlier law, which had said that courts "shall" use fees, to saying that courts "may" use fees but "only to the extent necessary" (205(e)), but this has had no apparent effect on the fees charged for use of PACER. Bills such as the *Open Courts Act of 2020*, which would have required replacing PACER with "one system" with full-text search functionality, available to the public free of charge and without requiring registration, have been adamantly opposed by the judiciary and have failed to become law (American Bar Association, 2020).

The adoption of CM/ECF and PACER by the federal judicial system resulted in a lot more information being more accessible. While "published" court opinions had always been widely available in law libraries, now those opinions were available through PACER as well. And all the other Public Information of the courts (unpublished opinions, dockets, briefs, testimony, and exhibits), which had only been available for in-person examination at individual court houses, was available online. Although the PACER system has many limitations on how case material can be found (Watson, 2010; Martin, 2008) and although it imposes fees for searching as well as for retrieving documents, it is undeniable that PACER has made more information more easily accessible than in the paper era.

But what about preservation? The *E-Government Act* and PACER were almost exclusively about "access." The *Act* did require courts to keep written opinions available online but only required other court documents to remain available online for one year after closing (205(b)(2)). Legally, all these documents are treated as Government Records, and their ultimate disposition (retention, discarding, or preservation) will be determined by the Records Schedules of the courts, and any access for any preserved Records will be provided by NARA. Although there is no complete inventory, it appears that, so far, most if not all courts are retaining court documents in their CM/ECF systems and continuing to make them accessible to the public through PACER.

Other changes have affected de-facto preservation. The 2005 decision to make "written opinions" available to the public without a fee opened the way for content aggregators to automatically harvest opinions for free (Martin, 2018). Google, for example, did this in 2009, making opinions available through Google Scholar (Acharya, 2009). There are, however, limitations to which "written opinions" are actually available for free. As noted above, the judge of each case determines whether or not a case fits the legal criteria for being a "written opinion." Only courts running a particular version of the CM/ECF software can participate. It was initially only available for District Courts. And cases decided before the release of the new software supporting this feature might not be available for free. Nevertheless, the work of the content aggregators resulted in more digital copies being stored and made available from more sources, and that gives them a better chance of being preserved through the principle of "lots of copies keeps stuff safe"—though without authentication, chain of custody, management to ensure against corruption or alteration, or guarantees of long-term access.

The big change to preservation began in 2010 when the Judicial Conference approved a one-year pilot project of working with GPO to provide "free public access to court opinions." In 2011, GPO received approval of the pilot project from the Joint Committee on Printing, began receiving digital transmissions from the courts, and began making them available on the web through FDsys, the precursor to govinfo.gov (Administrative Office of the US Courts, 2011). The technology behind the transfer was, again, a new version of the CM/ECF software, which pulled opinions (just "written opinions," not full case files or opinions that were not tagged as "written opinions") from the databases of the 30 participating courts and sent them to FDsys nightly along with metadata coded in XML (GPO, 2011b; Administrative Office of the US Courts, 2013a; Minick, 2011). Although the courts promoted this project as enhancing access, which it did, it had the additional benefit of putting these opinions into GPO's preservation repository, GOVINFO. The Judicial Conference approved the expansion of project in 2013, and it expanded to 64 courts. By 2018, the program had expanded to include 53 federal district courts, 50 bankruptcy courts, all 12 regional circuit courts of appeals, and the Court of International Trade (Martin, 2018).

Participation in this project is still voluntary for each court, and not all courts have chosen to participate. According to the "Browse by Author" page of govinfo.gov, as of mid-2022, GOVINFO has opinions for all 13 Courts of Appeal, 70 of the 94 district courts, 70 of the 94 bankruptcy courts, and one of the two special trial courts, the United States Court of

International Trade. The US Supreme Court is not included in this project. GOVINFO coverage of written court opinions comprises 75 percent (154 of 204) of the judicial branch federal courts.

This has produced a significant positive change in long-term preservation of government Public Information. It has also completely changed the character of what GPO is able to preserve. As mentioned earlier, only about one-tenth of 1 percent of the content GPO cataloged was from the judicial branch in the paper era. But, in the digital age, almost 75 percent of what GPO preserves in GOVINFO is from the judicial branch. The two graphs below compare the records in the *Catalog of Government Publications* by branch with the content in GOVINFO by branch. The barely visible line between the exec and legis sections of the graph 7.1 morphs into almost 75 percent of the content of GOVINFO in graph 7.2.

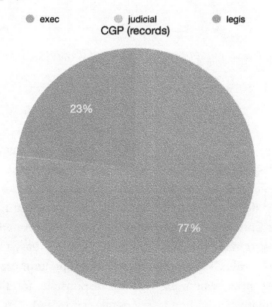

Graph 7.1. Percentage of records in the Catalog of Government Publications by branch.

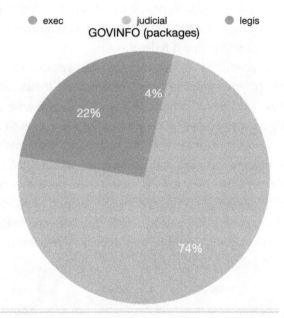

Graph 7.2. Percent of packages in GOVINFO, by branch.

Summary

This history of the shift from paper records to electronic records demonstrates how laws, technologies, and bureaucratic decisions can affect access and preservation both positively and negatively. The bureaucratic decision to manage court case files electronically opened the door for Congress to require that those records be made available on the web. It also made it technically and financially easy for the courts to share content with GPO for free access and long-term preservation. This demonstrates how, when the agency responsible for the creation of digital information is willing to cooperate with GPO, GPO can preserve its information.

It also shows how preservation based on such decisions are fragile; individual courts can simply fail to voluntarily deposit their information with GPO (25 percent of the courts have yet to join this voluntary project). It also suggests that when a producing agency profits from selling the information (as the courts do by using the PACER paywall), it has an incentive to protect its valuable information from being freely available (in this case, by withholding all the case files except the "written opinions" from GPO). These different treatments of short-term access to Public Information produce a patchwork of accessibility (Table 7.9). And this short-term accessibility translates into guaranteed long-term fee public access and preservation in only one instance: the written opinions that go into GOVINFO.

Type of Public Information	PACER	GOVINFO	commercial	web
traditional				y
written opinions	y	most	most	
unwritten opinions	y		some	
case files	y			

Table 7.9. Availability of judicial Public Information by type
of information and source.

For preservationists, these events and changes should raise many questions. For example, who will preserve judicial branch information that was once traditionally published and is now published direct to the web? What was the persuasive argument that led the judicial branch to decide to share written opinions with GPO? Will the courts or NARA preserve and provide long-term, web-based access to the digital court Records that are not going to GOVINFO? Will Congress pass laws requiring case files to go to GPO? Who will take responsibility for a comprehensive plan for preservation of born-digital judicial branch Public Information?

CHAPTER 8

Web Harvesting

In the next few chapters, we review the history and status of web harvesting by GPO, NARA, and the Library of Congress. We also compare two similar harvests: the End of Term harvest of 2020 (described in Chapter 6) and the content harvested by the Library of Congress in 2020. Our intent is to gather information that can be used to assess harvesting as a means of collecting and preserving in order to provide a better foundation for planning preservation actions.

Introduction

Web harvesting goes by many names (e.g., web capture, web archiving, web collecting). The process of harvesting itself is referred to alternately as capturing, collecting, extracting, replicating, scraping, spidering, or crawling. Just as these different terms reflect different aspects or goals, they also reveal that web harvesting encompasses a wide range of potential activities and results.

The Library of Congress defines web harvesting as "the process of collecting documents from the Internet and bringing them under local control for the purpose of preserving the documents in an archive" (US Library of Congress, 2007). Although harvests can be "manual" or "automatic," as a simple matter of scale most content that is preserved by web harvesting is automated. There are different approaches to harvesting web content (client-side, server-side, transactional) and different kinds of tools for harvesting (Pennock, 2013). The most common technique is the use of automated, client-side software that performs the collection of content under the constraint of rules that are configured by humans.

Automated web harvesting has become an accepted method of preserving web content. National libraries and archives started large-scale web-archiving projects as early as the mid-1990s. This included Australia's PANDORA and Sweden's Kulturarw3 in 1996 and the Nordic Web Archive in 2000. The largest project was the non-governmental Internet Archive, which Brewster Kahle launched in 1996. In 2003, 11 nations and the Internet Archive chartered the International Internet Preservation Consortium (IIPC), which has been instrumental in defining a standard architecture for web harvesting and in developing web-archiving tools (Pennock, 2013).

The process of web harvesting is conceptually simple: the software emulates a human visiting web pages, but it "harvests" (downloads and stores) the pages it visits, analyzes the links on those pages and visits those links ("crawls"), harvesting those pages and analyzing their links, and so on. Technically, it could continue this process indefinitely, but in practice, the human running the software configures it with rules limiting its reach. Limits can typically be placed on breadth (how far it will crawl away from the domains of the original starting points, called "seeds") and depth (how far into a domain or website it will continue recursively crawling).

Because of the many options involved in setting up an automated crawl, the outcomes of different harvests will be different. (For an example, see, below, the results of two test harvests that GPO conducted in 2006 of EPA websites.) Different software and different versions of the same software may have different rules, be configured differently at different times, and may perform differently. Harvests can be affected by how closely websites and web pages conform to web standards—standards that evolve over time. For example, a website may involve new techniques of database-driven display, navigation, and functionality, which web harvesting software was not designed to consider. This can result in unexpected crawling and unexpected harvesting. In general, harvesting software is intended to give humans who configure it control over what is collected, but in the end, it is the software that actually "decides" whether a particular piece of content on the web will be collected or not. The software's decision is based on its capabilities, the configuration options available to those who run the harvest, and how people configure those options.

CHAPTER 9
GPO Web Harvesting

Background

GPO's approach to government information on the web has gone through a variety of approaches over the years. Since GPO's inception, it has had two separate mandates: one to distribute publications to Federal Depository Libraries for access and preservation (44 USC Chapter 19) and one to create a comprehensive index and catalog of public documents (44 USC Chapter 17). In the paper era, these mandates overlapped and complemented each other, but when agencies posted documents on their websites and created new forms of publications, GPO's obligations became less clear. With the *GPO Access Act of 1993*, Congress gave GPO new mandates to "store" and provide online access to "appropriate publications" and to maintain an electronic directory of Federal electronic information (44 USC 4101). But this did not clarify GPO's role in preserving Public Information posted on agency websites.

GPO's approach to web-based content began with two assumptions that were based on its legal mandates. First, GPO decided that it must *avoid* acquiring, disseminating, cataloging, indexing, or preserving content that was not within the scope of Title 44 definitions—specifically, the infamous "publication... published as an individual document" limitation. Second, GPO's approach to its mandates for indexing and cataloging would be to catalog individual titles ("publications"), mostly using traditional library cataloging techniques. These two assumptions drove GPO's web harvesting activities for roughly 18 years (1993-2011).

Early Approaches to Web Content

GPO first addressed web-based content explicitly in 1995. It said that it would not attempt to put into the FDLP program (the Chapter 19 mandate) information that was "not previously available" (presumably including information that agencies posted to the web without sending copies to GPO) because it was not funded to do so. But it simultaneously decided that it would catalog (the Chapter 17 mandate) documents it found on the web (GPO, 1995b).

Less than a year later, it issued a strategic plan as part of its first major digital planning document, the *Study to identify measures necessary for a successful transition to a more electronic Federal Depository Library Program* (GPO, 1996b). The index and cataloging plan was to include URLs in the cataloging records of content on government websites, but not to provide "hot links" to the content in the online *Monthly Catalog*— which published the cataloging records. The logic GPO used at the time was that, because many people did not have internet access or even a computer, permanent access to cataloged government information would still be provided by paper copies in depository libraries, so "hot links" to websites were unnecessary even in the *Monthly Catalog*, which GPO had put online in June 1995.

Indexing If Not Archiving

But the *Study* did include a plan to complement traditional catalog records with an online index to web-based content. GPO would create an index that would provide users with a centralized mechanism for finding electronic government information products on multiple government web sites using "advanced indexing, search, and retrieval tools to identify, describe, and link users to electronic Government information, whether it is held by GPO or at other sites." GPO said that the *GPO Access Act* "required" it to create the index (possibly a reference to the requirement that GPO maintain an electronic directory of Federal electronic information). Despite this interpretation of the law, one of the *Study*'s task forces suggested amending sections 1710 and 1711 of Title 44 to explicitly put this complementary approach into law.

Manual Harvesting

These early strategies did not address how to preserve digital web-based information. GPO did, however, recognize the danger of losing information when agencies removed it from the web and the need for a mechanism or policy to ensure permanent public access to Public Information posted on agency websites. The 1996 *Study* suggested that GPO should "seek to establish arrangements" to ensure permanent access. It provided several case studies that outlined alternatives, but it did not propose a definitive answer. Web harvesting was not on GPO's agenda.

During the next few years, GPO used what it called manual and semi-manual harvesting as part of its regular workflow to add individual web-published documents to its catalog (GPO, 2007b). "Manual Harvesting" meant that a human would use a web browser to visit a URL and then save the content. "Semi-manual harvesting" meant that software was used to perform the steps of visiting a narrowly defined group of URLs, such as new issues of a serial publication, and saving the content. GPO saved copies of these harvested publications, but the number of documents acquired was small, and GPO's copies were not made available to the public.

In 2002, GPO began working with OCLC, which was developing tools for collecting, cataloging, and preserving web content. This was not an automated web-harvesting project but closer to GPO's own "semi-manual" harvesting. But it did use the "OCLC Digital Archive System" to store and preserve the harvested titles and make them publicly available (OCLC, 2003). The project rolled out in 2004 with 34 documents with PURLs (Persistent URLs) that pointed to the preserved objects (GPO, 2004d).

Automated Harvesting

By the early 2000s, GPO was aware that its manual harvesting was not keeping up with the great quantity of web-based content being produced by agencies. This was a "bottleneck" that was keeping content out of FDLP and the catalog, so GPO began taking steps to use automated web harvesting.

In 2004, GPO included web harvesting in the development plans for a new system to replace *GPO Access*. The replacement, dubbed the Future Digital System and later renamed the Federal Digital System (FDsys), was intended to be able to ingest, preserve, and provide access to digital information including web-harvested content. The documents that specified the concept and requirements of FDsys all had extensive specifications of an automated harvesting function that would "locate" content that was "in scope" from specified websites and then ingest, preserve, and deliver that content (GPO, 2004b; GPO, 2006b). GPO's 2004 "Strategic Vision for the 21st Century" did not mention web harvesting but did include a summary of the conceptual vision of FDsys including web harvesting (GPO, 2004c).

In December of 2004, GPO announced a pilot project to test automated web harvesting. A major focus of the project was to determine if web-harvesting software could distinguish between web content that was in scope of GPO's FDLP and Cataloging mandates (GPO, 2005a). GPO was so focused on finding software tools that would exclude out-of-scope documents that it cancelled its initial request for proposals, hoping that a "refreshed" RFP would attract more advanced software. Public Printer Bruce James, who had a background in printing, not computer information technology, apparently thought harvesting software would last for decades—the way printing machinery lasted. "We only want to do this once," he said. "[The question is] how do we get it right the first time, so we don't have to do it again 20 years from now" (Sternstein, 2005). In early 2006, GPO contracted with two vendors to conduct test harvests of publications from the Environmental Protection Agency's (EPA) web sites.

The two vendors each conducted three test crawls with harvester rules being refined for each new crawl. One vendor harvested about 400,000 documents of which 57 percent were identified as being in scope. The other vendor harvested almost two million documents of which only 5 percent were judged as in scope. GPO concluded that harvests could either concentrate on accuracy (getting a high percentage of in-scope documents) or comprehensiveness (getting more documents, but with a high percentage of content that was out of scope). GPO also realized that it did not have the resources to catalog harvested content. This test of one agency, EPA, had produced 200,000 documents to catalog—a task that GPO estimated would take it four years to complete. This just "moved the bottleneck" of getting web-based content into the program from the discovery and harvest functions into the classification and cataloging functions (Federal Depository Library Council, 2007).

In June of 2005, GPO announced a new policy for "Harvesting Federal Digital Publications for GPO's Information Dissemination (ID) Programs" (GPO, 2005b). The policy covered both manual and automated harvesting from federal agency web sites but was still limited to content that was within the scope of dissemination programs and focused on "final, published versions of agency publications." The policy specified that "out of scope files acquired via automated or manual harvesting will be deleted from GPO servers." By Fiscal Year 2005, GPO had harvested more than 6,000 documents and was at least anticipating the use of automated harvesting (GPO, 2006a).

As late as 2010, GPO was still listing automated harvesting of web content into FDsys as one of the three major steps to building a comprehensive collection of federal publications (Landgraf et al., 2010). The system was to be designed to determine what content to ingest based on whether or not it was in scope.

Policy Reversal

In 2008, GPO reversed its policy by updating its definition of what was in scope. It declared that content on publicly accessible government websites was "considered in scope" because it was on publicly accessible government websites (Schonfeld & Housewright, 2009a).

> All Federal information dissemination products published on an agency's (or an agency's official partner's) publicly accessible Web site and originating from or funded by the agency are intended for public use and are to be considered in scope for both the FDLP and C&I. (GPO, 2016d)

Four years later, GPO radically changed its harvesting activities. Instead of doing its own harvesting, GPO would outsource the job to the Internet Archive (GPO, Library Services and Content Management, 2011). Instead of trying to exclude web content that was out of scope, it operated under its new definition that declared web content in scope. Instead of trying to catalog one publication at a time, it would create a catalog record for an entire website, which GPO called a "collection." Instead of hosting harvested content in its own new online access system, FDsys, it would rely on the Internet Archive to host harvested content. Later, GPO said that, "as much as possible," it would migrate content that it had previously harvested and stored on its "permanent server" (presumably permanent.gpo.gov) to FDsys/govinfo. GPO also said that its long-term goal was to migrate content from the Internet Archive to FDsys/govinfo (GPO, 2016e).

GPO's contract with the Internet Archive service, *Archive-It*, created the Federal Depository Library Program Web Archive (FDLP-WA) (GPO, 2021a). This was GPO's first real automated web-harvesting project. It created "collections" based on websites. By 2014, GPO had archived 50 websites and, by 2016, 145. By late 2021, the FDLP-WA had 211 collections and 2,588 "seed" URLs. Each collection is for an office of the federal government (e.g., Bureau of Economic Analysis) or a program website (e.g., CrimeSolutions.gov) and includes several "seed" URLs, which provide the starting points for crawls. Typically, the list of seeds for a collection includes a primary domain (e.g., bea.gov) and subdomains (e.g., blog.bea.gov) and other resources such as social media (e.g., twitter.com/BEA_News/) (Braddock, 2021).

Evaluation

GPO does not claim that its web-harvesting program is comprehensive. Indeed, when asked about this by Congress, it responded that "GPO cannot keep pace with the scale of agency publishing directly to the web" and that it collaborates with other agencies "to harvest agency content and preserve this content on Library of Congress, NARA and other servers spread throughout the Government" (GPO, 2017). The number of websites that FDLP-WA harvested as of 2022 (211) is not even 1 percent of the number of websites discovered by the 2020 EOT harvest (31,103).

The original scope of the GPO harvesting project was "smaller sites" such as commissions, committees, and independent agencies, most of which are in the executive branch (Bower & Walls, 2014; FDLP, 2021a). We can get a better understanding of the scope of GPO's harvests by comparing the number of collections in FDLP-WA to the number of executive branch agencies and offices listed at USA.gov.

Preserving Government Information

category	agencies, offices, etc.	FDLP-WA collections	% of all agencies and offices
Executive Office of the President (EOP)	9	0	0%
Cabinet	15	0	0%
Departments, sub-agencies, and bureaus	260	34	13%
Independent Agencies	65	45	69%
Boards, commissions, committees	41	17	41%
Quasi-official agencies	11	2	18%
total	401	98	24%

Table 9.1. FDLP-WA coverage of executive branch agencies and offices.

Two things stand out in this table. First, overall, GPO is harvesting from less than a quarter of listed offices and agencies of the executive branch. Second, there are major gaps in GPO's coverage of the executive branch. FDLP-WA has no EOP or Cabinet-level websites at all and only 13 percent of major departments. GPO has targeted smaller websites and avoided larger ones. GPO harvests the "Millennium Challenge Corporation" but not the Office of Management and Budget or the departments of agriculture or commerce. It gets the "Armed Forces Retirement Home" but not the Army, Navy, or Air Force. It gets the Bureau of Justice Statistics, but not the Bureau of Labor Statistics. It gets small programs like Smokefree.gov, but not its host, the National Cancer

Institute. It misses most of the best-known government agencies like the CIA, BLS, DEA, FAA, FBI, FDA, NRC, NSA, OSHA, USGS, and the US Postal Service.

CHAPTER 10
NARA Web Harvesting

Introduction

No law explicitly requires the National Archives and Records Administration (NARA) to preserve government websites. But NARA's existing mandates and its interpretation of them have led it to harvest and maintain the websites of every Congress since the 108th (2004); it has also acquired every administration's whitehouse.gov website beginning with the first one (President Clinton's). In 2001, NARA required all federal agencies to harvest their own websites and deposit the harvests with NARA, and it administered its own harvest of all agencies in 2004. It later discarded the 2001 web harvests and announced that it would no longer harvest executive or judicial agency websites or require agencies to harvest their own.

Its several web-archiving projects differ in method, scope, frequency, legal authority, and justification across the three branches of government. This creates a patchwork of snapshots of the federal government web with large and conspicuous gaps.

The twisty history of web archiving at NARA is the result of its diligent efforts to meet a conflicting combination of technical and legal requirements and public expectations, and of finding ways to adapt paper-based archival traditions to digitally based records creation. It also conflates the differences between archival records and published Public Information and how both types will be preserved and made accessible. The reasons behind NARA's different decisions reveal how laws, technology, tradition, and the practical implementation of policy can both facilitate and hinder preservation. Understanding the difficulties of navigating this complex context will help digital preservationists plan for the future.

In this chapter, we describe NARA's web-preservation projects.

The President

NARA's oldest web-archiving program focuses on whitehouse.gov. It stretches back to 1995 with the first White House website, which was created in 1994 by the Clinton administration. Since then, NARA has consistently acquired the White House website at the end of each presidential administration. In some cases, NARA has acquired and preserved more than one snapshot. For example, on NARA's website of Archived Presidential White House Websites, there are six separate "archives" (essentially snapshots) of the Clinton White House, but only one of the George W. Bush White House (NARA, 2021a). In addition to single snapshots, NARA has also archived three special website collections for the Obama administration and six for the first Trump administration. Because it preserves whitehouse.gov only once per administration, the snapshots are either once every four years or once every eight years.

This web-archiving program is justified by a law originally enacted in 1978—before the world wide web was created.

As recently as the 1970s, at the time of the Watergate scandal when Congress wanted access to President Nixon's White House tapes, presidential records were legally the private property of the president. Congress passed the *Presidential Recordings and Materials Preservation Act* (1974) to prevent the destruction of the White House audio tapes, which, along with the president's papers, were then seized and housed in the National Archives. Four years later, the *Presidential Records Act of 1978* (PRA) gave the government the legal ownership of all presidential records beginning with President Reagan (Ferriero, 2017). The *Presidential and Federal Records Act Amendments of 2014* broadened the definition of the records covered by the law to include electronic information in "analog, digital, or any other form" (44 USC 2201).

Under current law, NARA takes both legal and physical custody of presidential paper and digital records when a president leaves office. Significantly, this includes the whitehouse.gov website. This means that NARA does not need to "harvest" the website; it simply takes control of it. On the day before the inauguration of a new president, whitehouse.gov is run by the White House. After the inauguration, NARA takes over operation of that old website using a new domain name (such as "trumpwhitehouse.archives.gov") and the new administration launches its own, new "whitehouse.gov" (NARA, 2008b). These websites are listed at https://www.archives.gov/presidential-records/research/archived-white-house-websites.

The legal, technical, and archival decisions that go into this act of preservation are complex.

NARA's justification for preserving whitehouse.gov is that "White House websites are Presidential records" (NARA, 2021a). Thus, NARA is simply following the law, which requires it to preserve all Presidential Records. This sounds simple and straightforward and an example of the law catching up with technology, but that is not completely accurate because parts of the White House website are not actually Presidential Records at all.

The White House website includes documents of the Executive Office of the President (EOP), and NARA preserves them in its "Executive Office of the President Electronic Records Archive." But, as NARA explains, the documents of some offices in the EOP are "federal records" and not "presidential records." Specifically, documents of the Office of Management and Budget, the Office of the United States Trade Representative, the Council on Environmental Quality, the Office of Science and Technology Policy, and the Office of National Drug Control Policy are considered "federal records" (NARA, 2017). These are covered, not by the *Presidential Records Act*, but by the *Federal Records Act* (FRA). Those laws treat preservation differently. All Presidential

Records are preserved, by law, but most Federal Records are discarded. Following NARA policy and guidance, most web-hosted federal records are considered temporary and are discarded. Nevertheless, NARA has decided to preserve all such records posted on whitehouse.gov.

It is rather easy to justify this decision either philosophically (it is good to preserve all White House Public Information) or as a matter of practicality (it is easier to preserve the whole website than to worry about how to discard parts of it while keeping its integrity). But such justifications have to be made in spite of the preservation laws, not because of those laws. It is also worth recognizing that preserving *all* presidential records, as required by law, varies from traditional archival practice of appraising records and preserving only a very small portion of them.

The intersection of the technology of accurately and selectively preserving web-based content, the laws governing what is supposed to be uniformly or selectively preserved, and archival traditions of appraisal did not provide an unambiguous solution to preserving whitehouse.gov. NARA had to navigate these uncharted waters and develop a practical approach, which it did.

Preservation of whitehouse.gov can be viewed as a case study that raises questions for preservation planning in the future. Recognizing the limitations of existing infrastructures of preservation is one step in planning for the future. Their roots in the paper era do not always translate effectively into the digital age. Clinging too closely to them can hinder successful planning.

Executive Agency and Judicial Branch Harvests

NARA has tried twice to preserve the public websites of executive and judicial branch agencies using web harvesting. It judged its first attempt in 2001 a failure, and it destroyed 20 terabytes of harvested data. After a second harvest in late 2004, it stopped harvesting agency websites altogether. As with the issue of preserving the presidential website, the harvesting of agency websites faced a complex intersection of laws, guidelines, technologies and traditions.

First Web Harvest, 2001

The first NARA web-harvesting project was in 2001 when the web was still in its relative infancy. CERN had released the World Wide Web into the public domain in 1993. When President Clinton took office that year, there were only about 500 web servers in the world, and most people had never heard of the web (United States Internet Council & International Technology and Trade Associates, 2000). The Clinton administration launched the first presidential website in 1994. By the end of the Clinton administration in January 2001, the government web had grown but was still relatively small (Graph 10.1).

Figure 2: Total .GOV Unique URLs

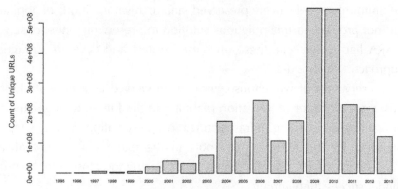

Graph 10.1. A graph of the number of unique URLs in the dot-gov webspace found in crawls by the Internet Archive from 1995–2013, constructed by Emily Gade and John Wilkerson. Spikes represent extra crawling during election years. (Gade & Wilkerson, 2017).

As the end of the Clinton administration in January 2001 approached, NARA was aware that agencies were publishing important information on the web and that not all "authentic records of policy" were being published on paper. NARA was also aware that agencies were producing new kinds of digital records on the web, records that had no parallel in the print world at all (NARA, 2001). But, at that time, no law required or authorized NARA to preserve agency websites or any web-based content. NARA had issued no regulations or guidance on preserving web-based content.

This was a transitional time for NARA's dealing with electronic information of all kinds, and existing policies and practices did not always result in clear answers to the questions of what should be preserved or how. For NARA, preservation is part of the life cycle of records management. In the final stage of records management, records are either destroyed or preserved. At the turn of the 21st century, many recordkeeping systems in federal agencies were still paper-based.

The issues NARA had to face with web harvesting in 2001 had parallels to its approach to email in the previous two decades. At that time, NARA did not consider emails as being Records. In the mid-1980s, even after National Security advisor John M. Poindexter and NSC staffer Oliver North deleted thousands of emails during the Iran/Contra affair, NARA still took the position that any email message that warranted being kept as a record must be printed out and stored in a proper recordkeeping system. Even if this was a good policy, it wasn't always fully implemented in practice. Agencies did not always provide an adequate infrastructure to support this policy. A court case revealed that there were only two printers at the NSC where more than 100 employees were regularly using email. At the end of the Reagan administration (1981–1989), it was expected that all its email would simply be erased to free disk space for the George H. W. Bush administration. It was only after two lawsuits challenged NARA's email policies that NARA changed those policies and classified email as "records" that must be scheduled for proper disposition (NARA, 1995; Wallace, 2001).

Perhaps influenced by its experience with email, by the turn of the century, NARA began to refer to web content as "records." This was not so much of a guarantee that web content would be preserved as a bureaucratic decision about records management. Traditional archival practice suggested that most Records are not worthy of preservation, but it at least requires Records to be considered for preservation.

NARA was updating its guidance on electronic records in the late 1990s and early 2000s but had yet to finalize any guidance on web records. Archivist John W. Carlin had established the Electronic Records Work Group in 1997 to review disposition of electronic information, and it issued its final report in 1998 (NARA, 1998). That report focused on creating procedures for scheduling email and other "office automation systems"—but not web content—as records (NARA. Electronic Records Work Group, 1998). In 2001, NARA was working with the San Diego Supercomputer Center and others on how to preserve "web records" (NARA, 2001).

In the absence of law or policies, NARA decided it had to act anyway—and quickly. It would use web harvesting to preserve "a one-time snapshot of agency public web sites as they exist on or before January 20, 2001" (inauguration day). Deputy archivist Lewis J. Bellardo said that any delay might result in a loss of records (GCN Staff, 2001). NARA later said that the decision was made out of "an abundance of caution" (NARA, 2008b).

It issued guidelines just eight days before George W. Bush was to take office, asking all federal agencies to "take a snapshot" of their public websites within the next eight days and send the snapshots to NARA. These snapshots would be "an archival record in the National Archives of the United States," which would "document at least in part agency use of the Internet at the end of the Clinton Administration" (NARA, 2001).

Agencies complained about this last-minute ultimatum. Web harvesting was a new concept for most, and they were not prepared or staffed for complying on such short notice. One agency even questioned NARA's authority to collect federal web pages, suggesting that web sites were not Records. Michael Miller, the director of NARA's Modern Records division, responded that most of the material on agency sites would qualify as records (Harris, 2001). NARA responded to agencies' misgivings by changing some of its original criteria for the quick harvest in order to make it easier for agencies to comply (GCN Staff, 2001). By mid-April, NARA had received 3,000 filings from agencies, comprising 20 terabytes of data (Powell, 2001).

The snapshots never went online, however. In fact, NARA later said that it was not certain that it could render the websites accurately for the public. By 2013, NARA referred to the data collected as "unscheduled records"—meaning that the records had never been properly described for disposition. Of course this was not news; no one ever claimed to have scheduled the records, and the harvests were done "out an abundance of caution" explicitly in spite of (or because of) this. NARA's examination of the snapshots determined that they had several problems: they were incomplete because of the "uneven rate of voluntary participation" by agencies, and the documentation submitted by the agencies varied widely in quality. It ultimately judged the attempt a "failed project" and destroyed the data collected (NARA, 2013).

It was not a complete failure, however, since NARA learned from the experience, and those lessons informed its eventual guidance for managing web records and its planning for its next web-harvesting project.

Second Web Harvest, 2004

Even before NARA announced its next web-harvesting project, it issued guidance to agencies that wanted to schedule web content records for long-term preservation, saying that "[w]eb content records are a priority electronic records format." These were rather primitive guidelines, however. NARA said, for example, that they did not cover "the wireless web," which it described as "[t]he use of the web through a wireless device, such as a cellular telephone or personal digital assistant (PDA)" (NARA, 2004a). The guidelines did improve the following year, however (NARA, 2005d).

NARA announced its second attempt at harvesting agency websites in October 2004. The harvest was to be a snapshot of the government web "on or before January 20, 2005," at the end of the first term of George W. Bush's administration. It had a new procedure for this harvest: NARA would perform the harvest rather than requiring agencies to do so—and it gave itself three months instead of eight days. It also had two new justifications. First, it said that it had "determined that a periodic 'snapshot' of Federal agency use of the Web provides important documentation of the Federal Government's activities in this area." Second, it said that one of the reports of the Interagency Committee on Government Information (ICGI), a committee created by the *E-Government Act* in 2002, recommended "a requirement for the capture and transfer of Federal public web sites at the end of an Administration" (NARA, 2004b; Interagency Committee on Government Information Web Content Management Workgroup, 2004). Of course, in October of 2004, no one knew if January of 2005 would be the "end of an administration" or the beginning of the second term of the Bush administration. But plans had to be made, and so they were. Neither of these justifications mentioned "records," but both presumed that preserving agency websites was important enough to justify doing so.

The harvest was conducted for NARA by the Internet Archive in late 2004 using a list of 982 civilian and military domains as seeds. It successfully harvested 75 million "web pages" from 1,370 civilian and military domains and 50,000 web sites. It obtained 6.5 terabytes of data, less than a third of the 20 terabytes harvested in 2001. It is likely that this harvest acquired so much less data than the 2001 harvest because the software was configured in 2004 to harvest no more than four levels of depth, making it less complete than the 2001 harvest (Kurtz, 2005; Theimer, 2008).

NARA made this harvest available to the public in January 2005 on its new webharvest.gov website as the "2004 Presidential Term Web Harvest (NARA, 2005b)." The Internet Archive designed, built, and maintains this website under contract with NARA (Internet Archive, [n.d.]). Text on the home page of the new website amplified the justifications for the harvest, saying that the ICGI didn't just recommend but "required" harvesting websites and not just at the end of an administration but at the end of each term (NARA, 2025). That wasn't what the law or the ICGI report said, but again, practical considerations apparently took precedence.

Decision to End Harvests of Agencies, 2008

Between the 2004 harvest and the end of the George W. Bush administration in January 2009, NARA settled on its approach to the content that agencies posted on the web. It would now follow the mandate that Congress gave it in the *E-Government Act of 2002*, which directed NARA to treat web content as Records under the *Federal Records Act*. This was part of an expanded view of electronic records, which was spelled out in a bulletin that described to agencies NARA's approach to electronic records in general, including web records (NARA, 2005a). It supplemented that bulletin with two publications specifically about web content: "NARA Guidance on Managing Web Records," its most comprehensive guide for agencies managing "web records" (NARA, 2005d), and "Implications of Recent Web Technologies for NARA Web Guidance," in which it described four technologies "that create content likely to exist only on the web" ("Web Portals, Really Simple Syndication (*RSS*), Web Logs (*Blogs*) and Wikis") (NARA, 2006). In 2008, it issued a new bulletin to remind agencies of "techniques that may assist them in scheduling existing electronic records" including "web records" (NARA, 2008d).

These were all part of a bigger shift in NARA policy from paper to digital. In 2005, after several years of research, it stated its vision for its "Electronic Records Archives" (ERA): to "authentically preserve and provide access to any kind of electronic record, free from dependency on any specific hardware or software, enabling NARA to carry out its mission into the future" (NARA, 2005c).

With these documents, NARA had produced a new infrastructure of regulations and guidance that unambiguously defined web content as Records that must be scheduled and managed as Records. Where practicality and concerns for lost information had taken precedence over law in 2004, now NARA would follow the law and tradition closely. NARA would attempt to apply the old traditions of appraising web content as Records.

As the end of the G.W. Bush administration approached, it was logical for NARA to apply this new approach to web content at the end of the Bush administration. It did so with a memorandum to agencies in March of 2008 that announced that NARA would not conduct a web harvest at the end of the Bush administration. In justifying this decision, NARA said that it had considered three things: the resources required for harvesting, NARA's other priorities for FY 2008, and the "availability of harvested web content at other 'archiving' sites (e.g., www.archive.org) [the Internet Archive]." The memo also mentioned that neither NARA nor agencies were required to harvest their websites, but that agencies were required to manage their web content as Records (NARA, 2008a).

This decision surprised and outraged many. A letter from the Project On Government Oversight (POGO), for example, complained about the decision and was signed by 20 organizations (including the American Library Association, the Association of Research Libraries, and the National Security Archive), three senators, and three members of the House. It asserted that "[a] snapshot of federal web pages is, indeed, of great historical value" and that failure to harvest the government web could result in the loss of "millions of pages" (Project On Government Oversight, 2008). This argument essentially mirrored the justifications that NARA had used in 2001 and 2004. The letter also noted that the work of "organizations like the Internet Archive, while valuable and meriting support, does not supplant the responsibility of our national government to protect and document its own history." This argument reiterated that oft-repeated "principle" that it was the government's responsibility to preserve its own digital information (see Chapter 1). In that light, it is perhaps ironic that NARA's decision to refrain from harvesting helped spur the creation of the End of Term Archive project to harvest the government web (Lippincott, 2018).

NARA responded to such complaints with a second memo, which explained that important "web records" would be preserved without a harvest by the same process that preserves other Records under the *Federal Records Act*. NARA also made it clear what this meant: "[M]ost web records do not warrant permanent retention and should be scheduled for disposal." The memo also criticized the very concept of web harvesting as a useful method of preservation:

> [I]t is not at all clear to NARA...that there is continuing
> permanent archival value of a Federal agency web snapshot
> taken on one random day... [T]he web snapshot does not
> systematically or completely document agency actions or
> functions in a meaningful way. Such records are found in
> other ongoing, systematic records series... (NARA, 2008b)

NARA had tried web harvesting twice and decided that its best course of action with regard to agency websites was to treat all web content as Records under the *FRA*. This decision was justifiable based on the law (the *FRA* and the *E-Government Act*) and on archival tradition of scheduling records and allowing agencies and NARA to decide what to preserve. This new policy was consistent with the law, archival tradition, budgetary priorities, and technology.

But this policy had an effect that went largely unmentioned and unexamined. It would treat published Public Information as Records. This was, effectively, a reversal of the way things had been. In the paper era, Records that agencies held but had not released could become Public Information by being scheduled for retention and transferring them to NARA. But, with this new decision, published Public Information—information that had been released to the public by posting it on government websites—would be classified as Records and most Records are discarded. In the absence of the deposit of published Public

Information into FDLP libraries, NARA would be the official preserver of this information, and the producing agency and NARA, not GPO policy or FDLP libraries, would decide what was worth saving.

We know that, in the paper era, "archival access" typically meant less access than "library access." But, in the digital age, we do not yet know what it means. We do know that virtually all new federal agency published Public Information (that is, web content on agency websites) would be preserved by NARA only if an agency captured that information in a recordkeeping system, scheduled it for long-term preservation, and successfully transferred it to NARA. We also know that NARA has advised agencies that most web records do not warrant permanent retention and should be scheduled for disposal. We know that if it web content is preserved, access to that information would be limited to whatever access NARA could afford to provide.

Preservationists will be wondering what the effects of this policy will be to long-term free public access to published Public Information. What will be preserved and who will make that decision? Will access to Records of published Public Information be different from access to other public but unpublished Records? Will access be adequate for the needs of a variety of user communities with different levels of technical needs and experience? To what extent will published Public Information be preserved and made accessible outside of NARA?

Congress

If NARA preserves the White House website because the *PRA* law compels it to do so, and preserves selected agency web content as Records because the *FRA* and *E-Government Act* direct it to do so, then what should it do when the law is silent about the preservation of Congressional websites? In spite of all the problems with web harvesting that NARA has identified, it has decided to harvest Congressional websites at the end of each Congress (NARA, 2008b).

In general, the law requires the work of Congress to be well documented and preserved. Laws written in the paper era are still active and include specific requirements for House and Senate documents and reports to be printed and delivered to NARA (e.g., 44 USC 7, 1714, and 2118). The Center for Legislative Archives at NARA contains historical records of the House and Senate and for legislative branch organizations such as the Congressional Budget Office and the Government Publishing Office (NARA, 2022c). An Advisory Committee on the Records of Congress reviews the management and preservation of the records of Congress (44 USC 27) (NARA, 2022b). The Secretary of the Senate and the Clerk of the House of Representatives are required to transfer all the noncurrent records of the Congress to NARA (44 USC 2118).

Unfortunately, the laws have not kept up with the digital shift. The *E-Government Act*, which requires agencies to treat their websites as Records under the *Federal Records Act*, does not apply to Congress (44 USC 2901(14)).

After harvesting congressional websites as part of the 2004 all-government harvest (see above), NARA began regular harvests of house.gov and senate.gov in 2006 during the closing weeks of the 109th Congress and has continued to harvest Congressional websites once every two years at the end of each Congress (NARA, 2021c). These are one-time, End of Term harvests and not ongoing harvests such as the Library of Congress's Congressional harvests (see Chapter 11). Coverage includes dot-gov websites of members, committees, leadership, and organizational offices (such as the Clerk of the House, but not legislative branch organizations such as the Congressional Budget Office). In recent years, it has expanded to include social media sites. The harvests are listed at https://www.archives.gov/legislative/research/web-harvest.html.

In announcing its decision to harvest congressional websites, NARA acknowledged the same shortcomings of web harvesting as a preservation tactic that it had enumerated as the reasons it would no longer harvest agency websites. It justified the decision to harvest congressional websites in spite of those shortcomings because Congress is not covered by the *FRA*. Once again, NARA said it was harvesting out of "an abundance of caution" and "to document" Congressional "presence on the World Wide Web" (NARA, 2008b).

Summary

NARA preserves all of whitehouse.gov under the *Presidential Records Act*. It preserves selected executive agency web content as Records under the *Federal Records Act* only to the extent that agencies either have that content in official recordkeeping systems or specifically schedule it for preservation. It harvests congressional websites because there is no law for the preservation of congressional websites. The table below summarizes the result of these decisions.

Congress (date)	Congressional (websites)	Presidential (whitehouse.gov)	executive & judicial except Supreme Court and PACER
103rd (1993-1994)			
104th (1995-1996)		Nov 1995 (Clinton)	
105th (1997-1998)			
106th (1999-2000)		1999, 2000 (Clinton)	[Jan 2001: deleted]
107th (2001-2002)			
108th (2003-2004)	late 2004 .gov crawl	late 2004 .gov crawl	late 2004 .gov crawl
109th (2005-2006)	2006 ("end of each congress")		
110th (2007-2008)	2008	January 2009 (Bush)	2008 announced: "no more"
111th (2009-2010)	2010		
112th (2011-2012)	2012		
113th (2013-2014)	2014		
114th (2015-2016)	2016	January 2017 (Obama)	
115th (2017-2018)	2018		

116th (2019-2020)	2020	January 2021 (Trump)	
117th (2021-2022)	2022		
118th (2023-2024)	2024	[January 2025 (Biden)]	

Table 10.1. NARA's website preservation, by branch and Congress.

CHAPTER 11
LC Web Harvesting

The Library of Congress's approach to web archiving of federal government Public Information is driven by a broad legal mandate, a long-term commitment to collecting information by and for Congress, flexible acquisitions policies, and a commitment to preservation.

As noted in Chapter 5, the Library of Congress's preservation mission is more general and open-ended than the very specific preservation mandates of GPO and NARA. Instead of having to hue closely to the specific wording of laws in order to determine what they are allowed—and, by inference, not allowed—to preserve, the Library has the flexibility to seek out information that needs to be preserved. So, while GPO, for example, spent more than 13 years (1995–2008) trying to conform its approach to web-based content to its Title 44 mandate (see Chapter 9), the Library of Congress was able to act more quickly and decisively and even retrospectively.

Thus, even though the law does not require LC to archive any websites, it has been actively doing so since 2000, and it has retrospectively acquired harvests back to 1996. Its justification for web archiving is clear:

> [T]he Library has been archiving born-digital online content
> through its Web Archiving program since 2000 in an effort to
> provide access to and preserve such materials as we have
> done with print materials throughout the Library's history.
> (US Library of Congress, [n.d.])

This commitment to web archiving in general, combined with the Library's long-time commitment to collecting and preserving government information (see Chapter 5) have led to a robust program of archiving government web content.

In this chapter, we analyze the content of the LC web archive of federal government Public Information.

The LC Web Archive

The Library of Congress has specific guidelines for web archiving that it spells out in its mission statement and its collection and preservation policies. It has a clearly articulated "Digital Strategy" that spells out its dedication to digital preservation, long-term stewardship of digital collections, and preservation of the utility of digital content (US Library of Congress, 2019).

Its collection policy for US government publications (US Library of Congress, 2016) includes publications in all formats including websites. But, while its general collection of federal government information is "comprehensive," its collection of federal government web content is not. It is comprehensive for the legislative and judicial branches, but only "selective" for the executive branch, the largest branch of the federal government. Its comprehensiveness in harvesting the legislative branch is exemplified by the fact that it harvests the congress.gov website, which the Library itself maintains based on data it gets from GPO (US Senate, Subcommittee of the Committee on Appropriation, 2019, p. 2).

While explicitly recognizing that "materials maintained on agency web sites routinely disappear," the Library says that it will have to share responsibility for web archiving and web preservation with GPO and NARA. The Library's reasoning is that must limit its harvesting because of "the large number and size of the Executive Branch websites" and can do so because of "the commitments by other agencies (GPO, NARA, etc.) to archive" (US Library of Congress, 2017).

By policy, the Library focuses on archiving cabinet-level agencies and affiliated programs that "complement the Library's Judicial and Legislative collections." It also archives websites of smaller agencies selectively. It does not harvest most dot-mil (military) sites or the web content of any of the national laboratories (of which there are more than a dozen, e.g., Frederick National Laboratory for Cancer Research, Sandia National Laboratory [USAgov, 2025]).

The Library of Congress Law Library has its own web archiving programs for federal courts and Congress (US Law Library of Congress, 2025), although these do not appear to be separate or distinct from the Library's programs.

The Library's approach to web harvesting is to "create an archival copy—essentially a snapshot—of how a site appeared" (Grotke, 2011). It contracts with the Internet Archive to do the harvesting. As of early 2023, the Library was still using the "OpenWayback" software to display the results on the Library's public website (US Library of Congress, [n.d.]). (According to its GitHub page [https://github.com/iipc/openwayback], OpenWayback is no longer under active development.)

The content of the LC Web Archive is defined by harvests it calls "archives." For example, see the web archive of the United States House of Representatives (US Library of Congress, [n.d.]). Each of these "archives" is defined by their seeds, which are the URLs where web harvest crawls begin, and "scopes," which define how far from those seeds the harvest is allowed to go.

For user browsing, these archives are grouped into "collections" on the LC web archive public website. These "collections" group several archives into a topic and are provided as a way for the public to browse the contents of the LC Web Archive. An "archive" may show up in many "collections." Each archive is defined by its seeds and scopes, not by "collections."

The Data

The Library of Congress generously created for us a large dataset of CDX ("crawler index") data, which included CDX records for dot-gov addresses harvested from 1996 through 2021. Just like the CDX data from the Internet Archive (IA) that we analyzed in Chapter 6, each CDX record provides data about a single visit of the harvester to a single URL.

The dataset consists of approximately three billion records from which we created two subsets for analysis. To create one subset ("LC-full"), we selected just those records of harvests from federal government hosts, using the lists of official federal government domains (described in Chapter 6). This created a subset of approximately two billion CDX records from federal government hosts by eliminating those from state and other non-federal dot-gov domains. We then created a second subset ("LC-2020") by extracting from LC-full just those 94 million CDX records of harvests during 2020.

1996–2021

Examining harvesting activity over time might help characterize a changing and evolving web as well as the activities and content of the archive itself. The number of records harvested each year for each branch, for example, may vary as much or more because of archival decisions by the harvesters than because of changes in what is being posted on (and withdrawn from) the web by the government.

As one might expect, the cumulative amount of dot-gov data archived by LC grew over time (Graph 11.1). Using CDX records as a surrogate for the size of the archive, we can see that the archive has grown from about three million items in 1996 to a total of two billion items by 2021.

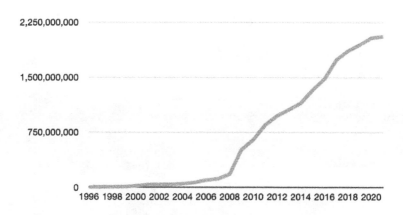

Graph 11.1. Cumulative number of federal dot-gov LC CDX records, 1996–2021.

Although the Library began archiving in 2000, it acquired, through purchase and sharing, harvests done by others before and after that. Notable among those acquisitions was the purchase from the Internet Archive of all their dot-gov harvests from 1996–2001, which added about 26 million records to the LC CDX dataset (Thomas, 2022a). The bulk of content archived, however, has been harvested since 2013 (US Library of Congress, 2017).

A count of CDX records by year of harvest shows increases in number of records harvested per year as harvesting activities grew, peaks and valleys as harvesting varied, and a tapering off in the last few years. The peaks of 2009 and 2017 in Graph 11.2 are apparently attributable to LC ingesting data for special projects: the 2016 EOT Crawl in 2017 and a California Digital Library project in 2009 (which was, presumably, the first End of Term Crawl).

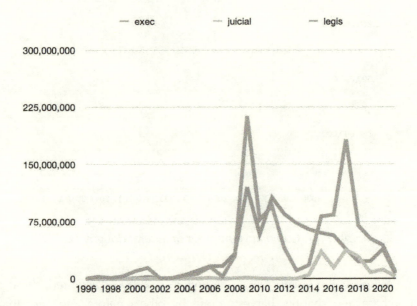

Graph 11.2. Number of federal LC web archive dot-gov
CDX records, 1996–2021, by branch, by year.

Although LC harvests the legislative and judicial branches comprehensively and the executive branch only selectively, the data harvested from the executive branch comprises more than 50 percent of LC's federal dot-gov web archives (Graph 11.3). The sheer size of the executive branch compared to the other branches explains this.

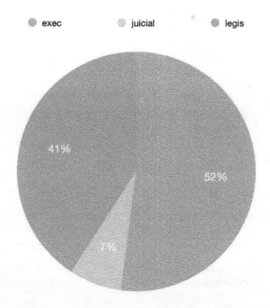

exec   juicial   legis

*Graph 11.3. Percentage of all federal dot-gov web records
archived by LC, 1996–2021, by branch.*

The number of CDX records is not a very precise way to measure the scope of harvests, however. It is not, for example, a count of URLs, but of the number of times harvesting software visited URLs. It is typical for the harvests to revisit URLs to see if a digital object has changed or moved, and each visit adds a CDX record. The number of CDX records is also not a count of "documents" or "titles." A single web page, for example, usually has many CDX records because there will be one for each component of that page (stylesheets, JavaScript files, images, etc.).

To provide an alternative view of the content of the LC web archive, we used counts of unique PDF files. Graph 11.4 shows number of PDFs by branch, by year harvested. The enormous increase in the number of PDFs in 2017 can be attributed to the Library's ingesting of 2016 EOT Crawl data. (One must wonder, however, why there was not a similar peak ingest of PDFs in 2009.)

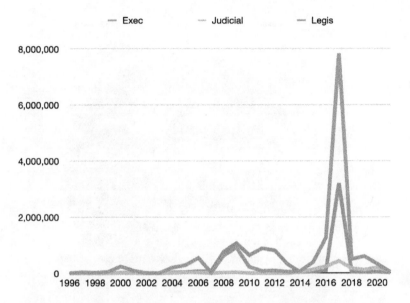

Graph 11.4. Number of PDFs archived from federal dot-gov websites by LC, 1996–2021, by branch, by year.

The total percentage of unique PDFs archived by branch provides a different perspective of LC's federal dot-gov web archives (Graph 11.5).

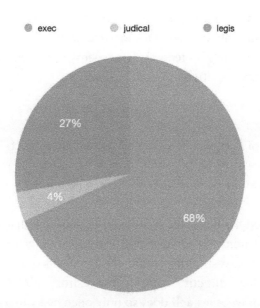

exec   judical   legis

27%

4%

68%

Graph 11.5. Percentage of unique PDFs, federal dot-gov LC
web archives, 1996–2021, by branch.

Table 11.1 provides data about 1996–2021 .gov information that LC
has archived. It lists, by branch, the cumulative totals of hosts, websites,
CDX records, HTML web pages, PDFs, and mime-types.

branch	hosts	websites	CDX (mils)	HTML (mils)	pdfs (mils)	mime types
exec	978	52,416	1,070	179	13	1,792
judic	8	1,472	147	35	0.7	197
legis	90	3,057	839	102	5	243

Table 11.1. LC .gov web archives, 1996–2021, by branch.

We believe these counts are reliable indicators (but not precise
measures) of the cumulative amount of Public Information posted on
federal government websites over 25 years. In a 2019 paper, Chase
Dooley and Grace Thomas of the Library of Congress found 19.2 million
PDF files in their examination of the complete LC dot-gov data,

including non-federal government websites. Our count of a total of 19.3 million federal-only unique PDFs in 2022 (with three additional years of data but excluding non-federal data) seems consistent with their count (Dooley & Thomas, 2019).

Summary

The Library of Congress has adapted new methods (web harvesting) to its traditional mission of selecting, acquiring, organizing, and preserving information for its communities of users, which include Congress and the nation.

It has used the flexibility of its legal mandates to harvest more public web content from the judicial and legislative branches than GPO and NARA. (GPO harvests almost none, though it gets quite a lot of data from both branches directly without harvesting. NARA only harvests congressional websites and does so only once every two years.)

CHAPTER 12
Comparing Two Harvests

Being able to compare harvesting done over 25 years with the harvests done in one year provides an opportunity to learn more about the government web and the harvesting and archiving of it. Having two different harvests of the same web space for the same year provides an opportunity to compare what they reveal about that web space.

Comparing 2020 with 1996–2021

We used our LC-2020 and LC-full datasets to compare what LC harvested in 2020 with what it archived between 1996 and 2021.

As one would expect, the LC-full dataset is much larger than LC-2020 since one dataset has 25 years of data and the other only one year. Table 12.1 compares the number of PDF files harvested in 2020 (about one half million) with the number harvested over 25 years (almost 25 million).

branch	PDFs LC-full	PDFs LC-2020	difference
exec	14,912,284	327,650	14,584,634
judic	1,510,056	199,388	1,310,668
legis	8,318,831	41,324	8,277,507
TOTAL	24,741,171	568,362	24,172,809

Table 12.1. PDF records in LC-Full and LC-2020 CDX datasets, by branch.

This does raise a question, however: Are those 25 million PDF files still on the web, or have they been withdrawn? If they are still on the web, shouldn't they have been re-harvested in 2020? Asking this question reveals the difficulty of using CDX data to characterize the government web. We cannot tell how much of the earlier archived data is still on the web from the CDX data because, when LC produced this datafile, it "de-duped" the data—meaning that data for an already harvested file was not included in the CDX dataset (Thomas, 2022b). Thus, the CDX data that LC provided only tells us about new files added each year.

With full analytical access to content in the web archive and to the full, original CDX data, researchers could identify content and classify it in one of four categories: content that A) has been withdrawn from the web entirely, B) has remained but been moved to a new location, C) has remained on the web unchanged, and D) has remained but that has been altered.

We did not have the access or the tools to do such a complete analysis, but we were able to do a small-scale analysis of categories A (withdrawn) and B (moved). We created a pseudo-random stratified list of 100 URLs of PDF files that were in the CDX data before 2020 but not in 2020. We then used the public interface to the LC web archive to see if those PDF files were still available in 2020 and examined those that were not. Only 15 of the 100 PDFs were available in the 2020 holdings of the web archive, leaving 85 as apparently withdrawn or moved.

Many of the URLs in our sample fit into category A, content withdrawn from the web. Many of these were non-current content such as press releases and announcements of meetings and lectures that happened decades ago. Other documents appeared to be parts of larger documents such as single page excerpts from the *Code of Federal Regulations* and the *Congressional Record*, a single slide from a presentation, a copy of one witness's written testimony from a congressional hearing. All of these are examples of content that agencies had, apparently, removed from the web—content that agencies might well regard as "out of date," or no longer relevant to the current business of the agency, or part of a larger document that was preserved elsewhere. From the posting agency's point of view, such files might well be judged as content that could be, even *should* be, removed from its website. In the words of NARA's advice to agencies for managing their "web records," information can be discarded "as soon as records are no longer needed" to conduct agency business (See Chapter 5).

We also examined documents that appeared to fall into category B, content that remained on the web but that had been moved to a new location. These included complete documents that had apparently been harvested from online "archives" (gao.gov, gpo.gov, and the Library's own *Chronicling America*) where one would expect documents to remain over time. We discovered that these were missing in the 2020 harvest because their addresses were no longer used—content had not been withdrawn but had been moved to a new address. Both the GAO and GPO archives had changed their URL naming conventions at least once over the years, creating link rot, but not missing documents. It is worth noting that, for these documents, the LC web archive successfully provides three important functions: it saves the documents themselves, it provides links to them using their original (now link-rotted) URL so that researchers with old references can still find them, and it provides a record of link rot that can be examined.

These preliminary findings suggest the need for more research to determine what is being posted, changed, moved, and removed from the public government webspace. This research will require the development of tools designed for this task, and archives will need to provide researchers with direct access to complete harvest records and the complete content of the archives. We believe that web archives should view providing such tools as an essential component of their responsibility.

Comparing LC Harvests With EOT

In Chapter 6, we used CDX data from the 2020 End of Term Archive crawl (EOT-2020) to characterize the size and scope of the government web. We can examine the accuracy of that understanding by comparing the EOT-2020 data with our LC-2020 data. The two datasets are similar in scope and composition but not identical, so comparisons are largely consistent and analogous but imprecise. One notable difference, for example, was that, because the LC dataset had been de-duped, we could not measure rates of duplication for the LC web archive as we did for the EOT archive.

As noted earlier (Chapters 8 and 9), it is normal for different harvests to acquire different content because of the specific choices made by those conducting the harvest. Such choices are based on scoping (what those conducting the harvest want to get and what they want to avoid getting), on implementation (how they tune their harvesting software to attempt to match their intended scope), and on execution (how the software interacts with each website). This means that comparing two different harvests of the same web space may reveal things about the web space itself, but it also reflects the differences in the scope and accuracy of the harvests themselves.

EOT-2020 was much larger (277 million records) than LC-2020 (94 million records) (Graph 12.1). Part of the difference in size may be attributed to the way the CDX files were created, which left out LC-2020 files already harvested in earlier years. The vast majority of those additional records in EOT-2020 came from the executive branch (Table 12.2). The Library's harvest contained only about 17 percent of the data that EOT did for the executive branch. It is likely that this difference is attributable to the different scopes of the harvests.

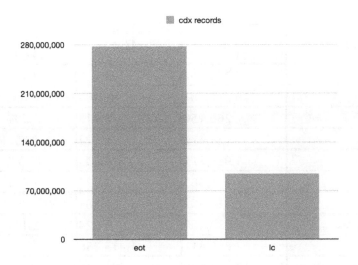

Graph 12.1 Number or 2020 CDX records for EOT-2020 and LC-2020

The scopes of the two harvests is revealed by an examination of the number of CDX records per executive branch host. LC had less than half the records that EOT did for the home web hosts of the cabinet offices of Agriculture (7 percent), Commerce (42 percent), Education (13 percent), Homeland Security (24 percent), Housing and Urban Development (1 percent), Justice (14 percent), and Treasury (1 percent). But LC had more CDX records for the cabinet offices of Defense (300 percent), Energy (129 percent), Health and Human Services (1,000 percent), Interior (3,000 percent), Labor (200 percent), State (800 percent), Transportation (400 percent), Veterans Affairs (1,000 percent).

Apparently, the scope of LC's harvesting is deeper for some executive branch areas than the scope of EOT's harvesting of those same areas. Conversely, LC's executive branch harvesting did not go as deep as the EOT harvest for some key sub-cabinet-level agencies (Table 12.2).

agency	EOT	LC	percent LC of EOT
Census	5,185,397	162,438	3.1%
EPA	16,227,381	99,303	0.6%
FCC	266,150	3,218	1.2%
FDIC	25,176	422	1.7%
FEMA	1,958,775	265,335	13.5%
FTC	155,585	2,125	1.4%
IRS	114,582	8,281	7.2%
Medicare	39,460	824	2.1%
NIH	7,855,417	1,415,256	18.0%
NOAA	7,193,125	4,097,336	57.0%
SEC	1,235,821	49,687	4.0%
SSA	86,099	1,392	1.6%
USGS	2,386,998	62,408	2.6%

Table 12.2. Number of 2020 EOT and LC CDX records for selected agencies.

In another contrast, LC acquired *more* data for the judicial and legislative branches than EOT did (Table 12.3).

harvest	exec	judicial	legis
EOT 2020	255,704,201	574,763	21,291,258
LC 2020	43,970,275	11,671,493	38,740,461

Table 12.3. Number of CDX records by branch harvested by the End of Term Archive 2020 crawl and by the Library of Congress in 2020.

As noted above, the number of CDX records is not a very precise way to measure the scope of harvests. Therefore, we also counted the number of web *sites* found by the different crawls (Table 12.4).

harvest	exec	judicial	legis
eot 2020	25,221	715	1,260
lc 2020	5,474	998	1,108

Table 12.4. Number of websites visited, by branch, by the
EOT-2020 and LC-2020 harvests.

LC's selective scope for the executive branch resulted in it harvesting data from only about 22 percent of the websites that EOT did. And, conversely, LC's comprehensive scope for the judiciary got it 40 percent more sites that EOT.

The only surprise here is that LC visited fewer legislative websites than EOT did. But the explanation for that may, again, be that the datasets we used were assembled in different ways, making comparison inexact.

The patterns are more complex when we count unique web *pages* (Table 12.5).

harvest	exec	judicial	legis
eot	102,663,329	285,672	16,144,558
lc	12,235,484	3,808,050	7,754,548

Table 12.5. Number of unique web pages by branch for the
2020 EOT and LC-2020 harvests.

Scope explains LC finding only 12 percent of the executive branch web pages that EOT found. In contrast, the LC CDX data for the judicial branch has more than 10 times as many web pages as the EOT CDX data has, probably because the LC's harvest was both "broader" (more seeds) and "deeper" (followed more links) than EOT harvest. In contrast to that, the EOT harvest of the legislative branch may have been deeper than LC's harvest since EOT found more than twice as many web pages as LC did from only a few more websites.

When counting unique PDFs, the numbers for the executive branch parallel the findings by websites and web pages as expected. But the number of PDFs found by LC are much lower than expected for the judicial and legislative branches (Table 12.6).

harvest	exec	judicial	legis
eot	1,928,759	166,236	686,321
lc	270,095	73,111	35,709

Table 12.6. Number of unique PDFs harvested, by branch, by EOT-2020 and LC-2020.

This unexpected discrepancy might be explained by the fact that, in creating the CDX data, LC de-duped the records. This would mean that PDFs that had been on judicial and legislative websites before 2020 were not listed as being harvested in 2020 and were, therefore, not included in our LC-2020 counts.

Summary

What can we conclude about the state of the government web from these two similar but very different harvests? Our estimates of the size of the government web in Chapter 6, which were based on the 2020 EOT crawl, were probably too small. We note particularly the order of magnitude differences between the EOT (285,672) and LC (3,808,050) harvests of judicial branch web pages.

Attempts to estimate the size of the government web using CDX data are hindered by the different ways the CDX data were compiled and the apparently uneven rates of harvesting from year to year. At best, CDX data can only provide an indirect indication of size, not a direct measurement of it. With that very large caveat, Table 12.7 uses the larger CDX counts from the EOT and LC 2020 harvests to create a very rough indicator of the one-year size of the "harvested" dot-gov web in 2020.

data type	exec	judicial	legis	totals
records	255,704,201	11,671,493	38,740,461	306,116,155
web pages	102,663,329	3,808,050	16,144,558	122,615,937
pdfs	1,928,759	166,236	686,321	2,781,316

Table 12.7. Total number of CDX records, unique web pages and unique PDF files found on the dot-gov web in 2020 by EOT-2020 and LC-2020.

CHAPTER 13
Gaps in Preservation

Although the government has done a lot to preserve its born-digital Public Information, there are still big gaps in what is being preserved, in the preservation infrastructure, in our knowledge and understanding of the issues, and, perhaps most importantly, in preservation planning. Our conclusions to Part Two are the identification of six gaps in preserving government information. These will provide starting points for preservation planning.

The Inventory Gap

We do not have an accurate, complete inventory of born-digital federal government Public Information. It is difficult to get even an accurate inventory of the sources of Public Information, i.e. the websites where they are published. Comparing the numbers of federal government websites we have found in our research (Table 13.1) illustrates how much disagreement there can be on such a basic concept.

branch	GSA	eot 2020	lc 2020
executive	10,018	25,221	5,474
judicial	9	715	998
legislative	65	1,260	1,108

Table 13.1. Numbers of federal government websites, by branch of government as counted by the General Services Administration, the 2020 EOT Crawl and our measurements of websites in the 2020 Library of Congress CDX data.

As for the published Public Information that is posted on public government web sites, we can use data from existing web harvests of those sites to get a preliminary characterization of the target of preservation. Since web harvests are incomplete, however, this is an incomplete inventory.

We estimate the size of the government web as relatively stable at around 1,800 large web hosts and many smaller websites. Well over 90 percent of the content of these web hosts is in two formats: web pages and PDF files. Using the EOT and LC harvests, we estimate that, in 2020, there were around 123 million web pages and around three million PDF files along with much smaller numbers of many other types of files (many of which are component parts of web pages).

branch	CDX records	web pages	PDFs
executive	255,704,201	102,663,329	1,928,759
judicial	11,671,493	3,808,050	166,236
legislative	38,740,461	16,144,558	686,321
TOTAL	306,116,155	122,615,937	2,781,316

Table 13.2. Estimate of the size of the government web in 2020 based on numbers of CDX records (all harvested content), web pages, and PDF files in EOT-2020 and LC-2020.

These estimates are rough indicators, not a direct measurement of the size of the government web. In addition to the simple gap of inventory of what is being created, we also do not know what existing Public Information is being altered, moved, and withdrawn. Without this information, we cannot know how successful we are at preserving what should be preserved.

The Preservation Gap

Our analysis of what is being preserved by NARA, GPO, and LC reveal some clear gaps in preservation. Most notable is that they are preserving only a very small percentage of the digital published Public Information produced by the executive branch, which is by far the biggest federal government publisher.

Web harvesting is quantitatively the largest method of preservation even though it is often inefficient and incomplete. Even those who use it seem to do so reluctantly and are careful to caution users that current web harvesting only provides an incomplete "snapshot" of how the web appeared on a random day, not a complete or meaningful record of an agency's publications.

Total preservation coverage (including harvesting and other methods) varies by branch of government, so we summarize each branch below.

Judicial

Public Information of the judicial branch seems to be largely well preserved. GPO's agreement with the Judicial Conference has had a positive result: most of the "written opinions" of 75 percent of federal courts are deposited into GPO's certified trusted digital repository, GOVINFO. The courts themselves are apparently preserving their own complete collection of digital Public Information in their own Case Management/Electronic Case Files (CM/ECF) system. That includes not only what they send to GPO but also what they do not send to GPO: opinions that do not qualify as being "written" and complete case files. That information is available to the public behind the paywalled PACER system. The long-established commercial market for judicial information provides a parallel system that, while neither free nor legally permanent nor complete, presumably includes more than the written opinions in GOVINFO.

The Library of Congress harvests judicial websites "comprehensively" and continuously, providing what we judge to be very good coverage. Indeed, it may have harvested more than 10 times as much data from the judicial branch in 2020 as the End of Term crawl did.

As good as it is, there are still gaps and risks of preservation of the judicial branch Public Information. The agreement between GPO and the Judicial Conference is, for example, incomplete in two ways. First, it is voluntary—so some courts do not participate and, presumably, any could back out if they wanted to. Second, it does not cover traditional publications, unwritten opinions, or case files at all. The preservation of that uncovered gap is inadequately defined in law. Although courts are obligated to transfer "permanent" content to NARA, exactly what it decides is permanent and what it decides to keep in its paywalled system is not yet clearly defined in law. We can only judge much of such content to be preserved provisionally and its permanence as risky.

Legislative

The legislative branch, including Congress, is, apparently, the most comprehensively and well-preserved branch. The main reason for this is that GPO is an office of Congress. Laws that require GPO involvement in congressional "publishing" result in Congress's long-time official publications being preserved in GOVINFO. For web-based publications, both LC and NARA do extensive web harvesting. LC harvests Congress "comprehensively" and continuously. NARA has been harvesting congressional websites at the end of each Congress since 2004.

Executive

As noted above, the executive branch produces the most Public Information by far but is the least well preserved. Very little of the executive branch's born-digital Public Information (only about 1 percent) goes into GPO's GOVINFO repository. This leaves web harvesting as the primary way that executive branch Public Information is being preserved by the government, and such harvesting is extremely incomplete.

While GPO, LC, and NARA explicitly recognize the need for cooperation in preservation and while each relies on the others to harvest what they do not, this cross-reliance is not coordinated enough to guarantee adequate coverage of the executive branch. LC says that it relies on "commitments by other agencies (GPO, NARA, etc.)" to archive the executive branch, but NARA has rejected harvesting as a means of preserving the executive branch Public Information, and GPO says that it relies on the Library of Congress, NARA, and other others to preserve web content.

As a result of these policies of relying on others, NARA does not harvest any executive branch content, GPO harvests from only 13 percent of major departments and does not harvest cabinet-level websites at all, and LC harvests executive branch content only selectively (leaving many large and important agencies and cabinet-level offices harvested at a rate of less than 10 percent).

The Legal gap

One of the main causes of gaps in preservation of government information can be traced to gaps in laws and policies. Although there is a general consensus that the government has an obligation to preserve its own information (see Chapter 1), the laws that affect preservation of born-digital Public Information are outdated and inadequate. Much of the law of digital preservation is little more than digital tack-ons to laws based on the paper-era infrastructure—an infrastructure that is no longer working. That infrastructure also relied on the assumption that preservation was a byproduct of access, and that assumption no longer applies in the digital age. New laws passed specifically to deal with digital content have failed to create a clear path to digital preservation of Public Information.

Policies to implement the laws sometimes help and sometimes hurt preservation and are sometimes just ignored. Policy implementation of laws seems to be based on a combination of how they are advantageous to the agency implementing the policy (i.e., save money, strengthen the agency politically) and good intentions. Sometimes this pattern helps preservation and sometimes hinders it.

The most successful government preservation projects so far (GPO's GOVINFO, LC's web harvesting, NARA's preservation of presidential websites) have relied on policies that preserve digital information in spite of the absence of specific legal requirements to do so. Relying on such policies is risky as a long-term strategy for two reasons. First, agencies can just as easily develop bad or inadequate policies (thus the *Chadha* problem described in Chapters 5 and 7). Second, policies are not as permanent as laws, and even good policies can easily be weakened or abandoned with a change in administration or a change in funding.

The Infrastructure Gap

In the paper era, preservation was a byproduct of laws and policies that focused on access to books. This access infrastructure, which deposited government documents into FDLP libraries, had the benefit of preserving those documents in hundreds of public, university, college, government, and law libraries throughout the country. It was not a perfect preservation infrastructure, but when it did work it worked well. Laws and regulations and practices converged into a clear path of both access to and long-term preservation of published Public Information. The access infrastructure was the preservation infrastructure. It ensured that published Public Information would not be altered or withdrawn or lost as long as communities of users wanted it.

The technologies of the digital age (digital publishing, the internet, and the world wide web) provide an infrastructure for information access, but not for preservation. There is no preservation infrastructure for born-digital Public Information that is comparable in scope or effectiveness to the infrastructure that existed in the print era. The digital access infrastructure provides no assurance that the information will remain unaltered or available for any length of time at all, much less for the long term.

Digital publishing has turned our understanding of access and preservation around 180 degrees. In the print era, we could rely on preservation being a byproduct of access because providing access more or less ensured preservation. But in the digital age, access stands alone without a preservation byproduct. It will be up to preservationists to build an infrastructure that makes access a byproduct of preservation (Conway, 1996).

The Preservation-Information Gap

To understand how successful existing preservation activities are, we need consistent standards for reporting and quantifying both the output of government publishing and the preservation of that output. Reports of terabytes of data preserved do not tell us what has been successfully preserved or where the gaps are. Similarly, using counts of inconsistently defined "web pages" or counts of URLs harvested are of little practical value in assessing the success of or gaps in preservation.

Born-digital information can be (and often is) moved, updated, altered, duplicated (presented in more than one way), and withdrawn completely from public access. We do not have reliable ways of measuring these changes or mapping such changes to what we want to preserve or what we are preserving. Link rot and content drift are real problems, and it is unclear to what extent web harvesting can successfully address these problems. It is likely that content worthy of preservation is being lost, but without proper accounting of what is being published and preserved, we cannot even know the extent of the loss, much less address it.

The Preservation Planning Gap

The gaps identified above suggest that there is one more gap: a gap in digital preservation planning. There are many projects to preserve government information, but too many of those tend only to "save stuff" rather than fulfill a comprehensive plan that addresses information discovery, delivery, readability, understandability, and usability of the full scope of government Public Information by Designated Communities for the long term.

The one project that comes closest to such a long-term planning strategy is GPO's GOVINFO repository. GPO's strategy included obtaining certification as an ISO 16363 trusted digital repository. That certification ensures that GPO is fulfilling and will continue to fulfill the six responsibilities and six functions of an OAIS repository (see Chapter 15). Although this is a practical plan for preserving digital content, it is not a comprehensive plan for preserving Public Information because of two problems.

The first problem in GOVINFO planning is a gap in acquisitions. The scope of content in GOVINFO is limited by law (44 USC Chapters 19 and 41), but the content actually acquired and preserved (as analyzed in Chapter 7) falls far short of even that limited scope. As noted in Chapter 5, GPO simply does not have the legal authority to enforce acquisitions of content it is legally allowed to preserve. GPO recognizes that it must rely on cooperation and coordination with others to preserve the full scope of Public Information (FDLP, 2021b), but this recognition is more aspirational than a plan of action. The result has been a continuing and growing preservation gap even of the limited scope specified by Title 44.

The second GOVINFO planning problem is its limited reach to users. The trusted digital repository certification of GOVINFO addresses the five guarantees of OAIS that information will be discoverable, deliverable, readable, understandable, and usable (see Chapter 4). But the degree to which these are adequate can only be measured in relationship to one or more Designated Communities. GPO's Designated Community for GOVINFO does not include the general public.

> The Designated Community for the system includes staff in Federal depository libraries, the United States Senate, the House of Representatives, the Administrative Office of the United States Courts, and the Office of the Federal Register. Members of the Designated Community are familiar with the organizations, documents, publications, and processes of the legislative, executive, and judicial branches of the United States Federal Government. The Designated Community is

able to access content information from the system and
render it electronically. PTAB's evaluation considers the
repository to be sufficiently usable for those who are
"familiar with the organizations, documents, publications,
and processes of the legislative, executive, and judicial
branches of the United States Federal Government" (PTAB
Primary Trustworthy Digital Repository Authorisation Body
Ltd, 2018).

A designated community of well-informed librarians and government
employees is a much, much smaller community that the "general public"
mandate of section 1911 of Title 44. It also contrasts with GPO's
promises to agencies that depositing their content with GPO will
guarantee that they will be "discoverable by a broad audience" and
"preserved in perpetuity for future generations of researchers, historians,
and the general public" (GPO, 2020c). It is much narrower than the
promise that GPO Director Hugh Halpern made to Congress in 2022
that important public documents are "made available to the general
public" through GOVINFO (Halpern, 2022). Although GPO consistently
advocates making information in GOVINFO "available" to the general
public, it does not choose the general public as its Designated
Community.

Together, these two problems of GOVINFO highlight the gap in a
comprehensive long-term strategy for preserving Public Information. The
gap in preservation planning is also evident from the imprecise popular
reasons promoted for preserving government information (Chapter 1), the
loose standards of "preservation" used to justify existing preservation
projects (Chapter 4), and the existing gaps in preservation (Part Two).

Conclusions

Although lots of good work has saved a lot of government information
that might have otherwise been lost forever, much work remains to be

done. These six gaps provide a starting place for thinking about what needs to be done and a template for evaluating future actions.

These gaps also raise many questions. How can government best address its preservation responsibilities? How can laws ensure digital preservation and avoid hindering it? What would a digital preservation infrastructure look like? What would a preservation plan look like? Such questions provide a framework for thinking about a digital preservation infrastructure in Part Three.

PART THREE
Preservation Infrastructures

One of the key obstacles to digital preservation is the absence of a digital preservation infrastructure.

In Parts One and Two, we repeatedly referred to the "infrastructure" that ensured the preservation of government information in the print era and the absence of a preservation infrastructure for the digital age. In Part Three, we address the foundations of an infrastructure for preserving government information.

We begin by defining "infrastructure" (Chapter 14) and introducing some important OAIS concepts and terminology (Chapter 15). We then describe how the inherent characteristics of print and digital information affect their preservation (Chapter 16). With that as context, we examine how this affects traditional assumptions about preservation, including the roles of libraries and archives and introduce some elements needed for a digital information preservation infrastructure (Chapter 17).

CHAPTER 14
Infrastructures

What Are Infrastructures?

What do we talk about when we talk about infrastructures? Apparently, a lot of different things! In its most common usage, "infrastructure" is used to mean things like roads, bridges, telephone wires, and network cables. These are the physical things, the "substrate," on which big systems like transportation and communication run. In the late 20th and early 21st century, for example, there was much written about a "cyberinfrastructure" that was equated with an internet that was fast and reliable. One dictionary points out that "*Infra-* means 'below;' so the infrastructure is the 'underlying structure' of a country and its economy, the *fixed installations* that it needs in order *to function*" [emphasis added].

But the word has also become a generic term for all kinds of complex systems and functions that the substrate supports. Christine Borgman, Distinguished Professor and Presidential Chair in Information Studies at UCLA, for example, writes about scholarly infrastructures (Borgman, 2007b), and Richard Cox, professor of archival studies, writing about the future of archives, describes documentary infrastructures (Cox, 2013). Ed Summers, a software developer who writes about technology at the intersection of libraries and archives and the World Wide Web and the ethnography of infrastructures, mentions the scholarly communication infrastructure, the space science infrastructure, data infrastructures, information technology infrastructures, repository infrastructures, web server infrastructures, network infrastructures, the web archiving infrastructure, and the infrastructure of the web (Summers, 2020). And, of course, there is a literature about a "preservation infrastructure" for information (National Digital Information Infrastructure and Preservation Program [US], 2002; Granger, 2002; Skinner & Halbert, 2009). Another dictionary recognizes this equating of infrastructures with entire systems and not just the substrates supporting them, saying that infrastructures are "social" and "economic" and even "organizational structures."

In stark contrast to these big-picture infrastructures, some technologists describe small, individual technologies as infrastructures. For example, Matt Burton, in a dissertation for a PhD in information, calls "blogs" an infrastructure for scholarly communication (Burton, 2015), and Clifford Lynch, the director of the Coalition for Networked Information (CNI), wrote about institutional repositories as the "Essential Infrastructure for Scholarship in the Digital Age" (Lynch, 2003). Jeremy York of the HathiTrust describes the HathiTrust digital library as an infrastructure. Borgman describes libraries as part of a nation's information infrastructure (Borgman, 2003). The Catalog of Open Infrastructure Services lists services such as DOI persistent identifiers, Jupyter Notebooks, and ORCID researcher IDs as infrastructures (Invest in Open Infrastructure, 2022).

These very different kinds of "infrastructures" overlap each other vertically. A discussion of a "transportation infrastructure" is made up of smaller infrastructures (e.g., road infrastructure, rail infrastructure, air-transport infrastructure). They also overlap horizontally with related infrastructures (e.g., vehicles infrastructure, standards infrastructures used to create vehicles, legal infrastructures that regulate their use, and so on). A broad description of a "housing infrastructure" would not just include houses and apartment buildings, but all the other physical and social infrastructures that make houses and apartments livable: energy, water and sewage infrastructures, transportation infrastructures, schools and parks, building codes, zoning laws, etc.

These very different infrastructures are often named but not precisely defined. These names are often used almost as metaphors for a hard-to-define mix of hardware, software, legal policies, formal and informal standards, social customs, culture, education, and communities that work together to accomplish some societal function in ways that are often hard to define. A digital preservation infrastructure might well include institutional repositories and HathiTrust, which might, in turn, encompass the software they use to preserve digital content. But it would also include much more.

When Are Infrastructures?

Although it may seem that it is impossible to have a single clear definition of "infrastructure," one thing is clear. In these more modern uses of the term, infrastructures are much more than isolated, built, physical things like roads and bridges and other substrates. Today, we use the word infrastructure as a way of describing complex systems that interact for a functional purpose. We use the word infrastructure when we want to describe and understand why some big-picture social function works—or fails to work. When we start talking about an infrastructure, we do so to explain what the pieces are and how (and why) they fit together. Although an infrastructure may include physical things that we build (like roads and bridges), an infrastructure is no longer just a thing that we can simply build.

This resonates with our findings in Part Two. Preserving government information is not so much a technical task as a social one. The preservation gaps identified in Part Two are caused by gaps in information, gaps in laws, and gaps in planning, not gaps in technologies. These gaps will not be filled by individual institutions or isolated projects.

Infrastructures are sets of relationships. As sociologist Susan Leigh Star and computer and information scientist Karen Ruhleder said, in their highly influential and much-cited 1996 paper "Steps toward an Ecology of Infrastructure," infrastructure is not the substrate, the thing upon which something else runs like a system of railroad tracks upon which rail cars run. Rather, they say, "infrastructure is something that emerges for people in practice, connected to activities and structures." This leads them to ask, not "what is an infrastructure?" but *"when* is an infrastructure?" (Star & Ruhleder, 1996) They ask not what is the substrate, but when do the relationships emerge?

For our purposes, the Star and Ruhleder understanding of infrastructures is the best model for understanding how information preservation infrastructures succeed or fail. They identified eight characteristics of infrastructures and Star and informatics professor Geoffrey Bowker later added a ninth characteristic to the list (Bowker & Star, 1999).

- Embeddedness. Infrastructure is "sunk" into, inside of, other structures, social arrangements, and technologies;
- Transparency. Infrastructure is transparent to use, in the sense that it does not have to be reinvented each time or assembled for each task, but invisibly supports those tasks;
- Reach or scope. This may be either spatial or temporal— infrastructure has reach beyond a single event or one-site practice;
- Learned as part of membership. The taken-for-grantedness of artifacts and organizational arrangements is a sine qua non of membership in a community of practice. Strangers and outsiders encounter infrastructure as a target object to be learned about. New participants acquire a naturalized familiarity with its objects as they become members;
- Links with conventions of practice. Infrastructure both shapes and is shaped by the conventions of a community of practice, e.g. the ways that cycles of day-night work are affected by and affect electrical power rates and needs. Generations of typists have learned the QWERTY keyboard; its limitations are inherited by the computer keyboard and thence by the design of today's computer furniture;
- Embodiment of standards. Modified by scope and often by conflicting conventions, infrastructure takes on transparency by

plugging into other infrastructures and tools in a standardized fashion;

- Built on an installed base. Infrastructure does not grow de novo; it wrestles with the "inertia of the installed base" and inherits strengths and limitations from that base. Optical fibers run along old railroad lines; new systems are designed for backward compatibility; and failing to account for these constraints may be fatal or distorting to new development processes;

- Becomes visible upon breakdown. The normally invisible quality of working infrastructure becomes visible when it breaks: the server is down, the bridge washes out, there is a power blackout. Even when there are backup mechanisms or procedures, their existence further highlights the now-visible infrastructure;

- Infrastructure is fixed in modular increments, not all at once or globally. Because infrastructure is big, layered, and complex, and because it means different things locally, it is never changed from above. Changes take time and negotiation, and adjustment with other aspects of the systems involved.

A Digital Preservation Infrastructure (DPI) with these characteristics might be called a *culture of preservation*. Thinking of an infrastructure in this way will help us envision a DPI that will succeed in the digital age. When we use the term "infrastructure" in this book, we are referring to something that has these nine characteristics. This will help us understand how the preservation infrastructure of the print era successfully preserved government information and why there are preservation gaps in the digital age.

The Print-Era Access Infrastructure

When a government agency wanted to make information public in the print era, it would publish (i.e., print) the information as a book, a map, a pamphlet, or other physical document. But the mere act of publishing did not automatically make the information easily accessible. To be usable, a publication had to be physically transported to each user. Many copies were needed in order to make them available to more than one person at a time. This inherent characteristic of printed publications led to the use of libraries as part of the information infrastructure. Libraries were ideal places for storing copies near users, and the Federal Depository Library Program (FDLP) provided more than 1,000 libraries for this purpose. The FDLP law (Title 44, Chapter 19 of the US Code) ensured these were geographically dispersed. Libraries were already equipped with staff, procedures, rules, standards, shared norms, policies, and practices that enabled them to easily select, receive, process, describe, store, and provide access to and services for government documents. In an era when one of the main metrics for measuring the quality of libraries was collection size, libraries benefited from receiving government documents on deposit from the government because it increased their volume count. The government centralized its printing for efficiency and ran the depository program from GPO, the Government *Printing* Office. People who needed government information could go to their local library and use familiar tools (human reference services, card catalogs, printed indexes and, later, OPACs and databases) to locate government documents the same way they located and used novels and encyclopedias. When a library became an FDLP library, it agreed that the general public could use its deposited government documents, even if the library was not otherwise open to the public.

This infrastructure was designed to provide access to printed publications, and it fit all of Star and Ruhleder's categories. The publishing of documents and distribution to libraries for use by the public was *embedded* in physical, social, and technological structures. GPO and libraries and users were all able to use this system of access *transparently*, and the *scope* of government information was virtually universal—no need to for special tasks with each new publication or type of publication. Publishers and librarians were *members* of professions where they learned practices and procedures and *conventions*. Publishing and catalog *standards* ensured the *transparency* and *scope*. The depository system was built on the *installed base* of publishing and distribution to communities of public, law, university, and special libraries.

Although this infrastructure was designed for access, it provided preservation as well. Because the traditional *temporal scope* of libraries was long term, this infrastructure provided *long-term* access. Providing access in the future is one definition of "preservation," so the infrastructure designed for short-term access had the byproduct of providing preservation. An update to Title 44 in 1962 instantiated this convention into law by requiring certain depository libraries ("regional depositories") to acquire all available depository publications and retain them permanently (44 USC 1911-1912).

Star and Ruhleder's categories also help explain the preservation problems introduced with the shift to digital publishing. Both the access infrastructure of depository libraries and its preservation byproduct *broke down and became visible* when documents were no longer printed or deposited with libraries. The inherent characteristics of digital publications did not fit easily into the embedded infrastructure that was designed for printed publications. GPO could have chosen to adapt the existing paper depository model. GPO could have, for example, simply sent digital copies of publication to Federal Depository Libraries (FDLs) (Jacobs & Jacobs, 2019). The transmission of digital documents would have been much less expensive than the shipping of paper documents. But neither GPO nor FDLs had either the substrate of hardware and software or the full infrastructure for sending, receiving, and processing digital publications (MacGilvray & Walters, 1995). There was no digital depository infrastructure in place, and to adopt digital deposit at the time would have required major changes by both GPO and FDLs. FDLP libraries would have needed to develop a new infrastructure to handle digital publications, digital access, digital services, and digital preservation. That would have required not just hardware and software (substrate) but new *training, policies,* and a new *culture* of digital access, digital services, and digital preservation—all of which were lacking in most FDLP libraries in the 1990s.

And GPO would have needed more than the (substrate) ability to send ("push") digital files or make digital files available for libraries to download ("pull") at their convenience. It would have also needed to develop policies and procedures for identifying which files to make available in which formats for which libraries. Even the concepts of "documents or publications" do not map clearly and consistently to digital concepts like "files," and GPO would have needed to construct new social understandings of what those meant for selection and service. GPO would have needed new standards, new conventions of practice, and so forth. For a variety of reasons, in the mid-1990s, neither GPO nor the library community developed a new infrastructure that relied on libraries to provide access and services and preservation.

What GPO was willing to do, and did do, was switch to a centralized, top-down approach that explicitly replaced decentralized deposit (and reliance on libraries) with centralized storage and online "access." This replaced reliance on libraries with reliance on GPO. GPO developed the infrastructure it needed for centralized storage and access. FDLP libraries largely accepted this solution because of their lack of digital infrastructure.

Both the publishing of government information on the web and this new GPO infrastructure were a replacement for the old access infrastructure with a new infrastructure that also provided access. In fact, in many ways, it provided better access than the old infrastructure by making government information instantly available over the web to anyone anywhere. But the new access infrastructure did not have a preservation byproduct. Preservation was *embedded* and *invisible* in the old access infrastructure, but it was missing in the new infrastructure.

GPO's choice to take sole responsibility for access (even long-term access) overlooked the need for *modular, local incremental* adjustment and use of an *installed base*. The result was the big gaps in preservation that we saw in Chapter 7. As Stewart Granger of the University of Leeds noted as recently as 2002, we need a "collaborative mechanism" to develop a deep infrastructure for digital preservation, "but such collaborative mechanisms do not currently exist" (Granger, 2002).

A Digital Preservation Infrastructure (DPI)

If, as Star and Ruhleder say, infrastructure is not a thing that we build but a set of relationships between people that emerges through practice, activities, and structures, then preservationists need two things to move forward: a framework within which preservationists can pursue actions that fit a comprehensive preservation plan and practical steps that we can take now. Before introducing a framework and actions, we can begin by describing the characteristics of a successful digital preservation infrastructure using Star and Ruhleder's terminology.

A DPI will build on an *installed base*, which means it will *inherit the strengths* that already exist, and it will avoid trying to create a new system from scratch (*de novo*). It will also be *modular* and will, by nature, be expected to grow, evolve, and adapt over time. It will *not be imposed from above* and it will *mean different things locally*. There will be *standards and conventions* and *organized practices* that will not be imposed but, when used, will enable libraries to provide better services and build collections more efficiently and effectively. The *standards will not be rigid* but flexible. They *will not rely on a lowest common denominator* (one size fits all) but will embody flexibility that can *grow* and evolve to *adapt to change*: changing forms of information, changing groups of users, and changing uses. All these will have to be acceptable to *different communities of practice* (librarians, archivists, publishers) so that they become *embedded* in social arrangements and technologies and part of accepted social norms. They will be *transparent* because, in the long run, they will be *learned as part of professional education* and *utilized in daily practice*, not tacked on as a special service isolated from normal practices. They will make preservation happen *easily and transparently* and *without the need to reinvent* practices for each new task or document or file-type or website.

From Infrastructure to Ecosystem

The title of the Star and Ruhleder article ("Steps Toward an Ecology of Infrastructure") is a hat tip to the famous book of essays *Steps to an Ecology of Mind* by Gregory Bateson (Bateson, 1972). Bateson trained as a biologist, but his career spanned anthropology, linguistics, psychology, and cybernetics. He was writing about schizophrenia when he wrote in one of those essays that "[w]hat can be studied is always a relationship or an infinite regress of relationships. Never a 'thing.'"

Star and Ruhleder saw how this analysis applied equally well to "the delicate balance of language and practice across communities and parts of organizations." Essentially, they argued that an infrastructure is like an ecology or a biological ecosystem in which organisms relate to one another and to their environment. When we think of a digital preservation infrastructure, we can understand it best by understanding how its constituent parts interact much the way biologists understand how the different parts of a successful ecosystem balance organically in order to thrive. As Star and Ruhleder said, using this analogy to ecosystems "draws attention to that balance (or lack of it)" of an infrastructure.

If we are to have a successfully functioning digital preservation infrastructure, we will need more than some technological tools, or rules imposed from above, or projects that are isolated from common practice. We will need to imagine a balanced ecology of preservation happening naturally, organically.

This is not a new idea. In the early days of the digital shift, the Commission on Preservation and Access and the Research Libraries Group created a Task Force on Archiving of Digital Information. They concluded that digital preservation was not a technical problem but "a grander problem of organizing ourselves over time and as a society to maneuver effectively in a digital landscape" (Waters & Garrett, 1996).

CHAPTER 15
OAIS

Envisioning the Invisible

As Star and Ruhleder point out, a working infrastructure is invisible; it supports tasks so well that we don't think about it (Star & Ruhleder, 1996). The infrastructure of the print era supported both access and preservation this way but failed to deal successfully with digital publishing. That means that preservationists in the digital age must envision a new infrastructure that will support digital objects. A good place to start that process of envisioning is by using preservation terminology consistently to define the functions of preservation.

This chapter provides an overview of the essential functions and terminology associated with the preservation of digital information. Digital preservation is a very large and often complex topic, and the preservation of Public Information is a sub-topic that contains its own unique issues. But in this chapter, we focus on the functional issues of digital preservation in general that do not vary and are not affected by format or type of content or technologies or time. Understanding these functions will provide the terminology that should inform any discussion of preservation, including digital preservation.

The Reference Model for an Open Archival Information System, OAIS, was developed to provide a common set of terms and concepts to help us understand and compare organizations that preserve information for access and use. It has been adopted as ISO standard 14721 (ISO, 2012b) and is, effectively, the standard for digital preservation.

What OAIS Is—And Is Not

OAIS describes functions but does not prescribe how to implement those functions. OAIS is a "recommended practice" document that describes a "reference model" for organizations that have a mandate to preserve information for the long term. A "reference model" is a very specific kind of standard that describes concepts and terminology to help practitioners understand, describe, evaluate, and compare systems. It does not prescribe how to do something, but how something should function. If there were a reference model for a "land vehicle," it might require functions such as propulsion and steering and stopping. Such a model could be used in the design of vehicles that implemented those functions in very different ways with vastly different costs, speeds, capacities, and efficiencies. It could be used to design a bicycle or motorcycle, a van or a bus, a pickup truck or a tractor-trailer, a gas-powered automobile or an electric vehicle, a pogo stick or a *Star Wars* "landspeeder."

OAIS does not prescribe any particular methods of designing, implementing, or managing an archive. It does not require specific software, hardware, file formats, database schemas, or metadata standards. Nor does it demand any technological solutions. Indeed, although its focus is on digital information, it is so generalizable that it is applicable to non-digital information as well. It is not about technology; it is about preservation of information. And information can include "a sequence of bits, a table of numbers, the characters on a page, the recording of sounds made by a person speaking, or a moon rock specimen."

OAIS does not give answers—it prompts an archive to ask questions. By focusing on functions, it prompts preservationists to ask questions such as: Who are we doing this for? What information will we preserve? How will we ensure that information will be understandable years from now? Different organizations will answer these questions differently, and their answers will help define their collections and services.

OAIS is also the basis of the certification standard called *Audit and Certification of Trustworthy Digital Repositories* (ISO 16363 also sometimes referred to as "TDR") (ISO, 2012a). Although institutions seeking certification as Trusted Digital Repositories have to understand OAIS, preservation institutions can use OAIS to design and describe their archive without using TDR or taking the extra steps needed for certification. The TDR standard is designed to test how closely a particular institution conforms to the OAIS standard. For example, GPO applied for and obtained TDR certification for its GOVINFO repository (PTAB Primary Trustworthy Digital Repository Authorisation Body Ltd, 2018). The certification process can be lengthy and expensive for institutions that adopt OAIS principles late in the process of preservation planning. But OAIS is useful to all preservation planning whether or not the institution using it desires to get TDR certification. Using OAIS for planning does not have to be expensive or time consuming, and using it thoughtfully during the early planning process should have more benefits than costs in the long term. (See also Appendix A.)

OAIS gives preservationists a consistent, unambiguous vocabulary for discussing, designing, describing, managing, and evaluating digital repositories of all sizes and kinds. It enumerates the essential responsibilities of preservation, enabling preservationists to ensure their planning is complete.

The OAIS standard defines a reference model for a "system" (an "open archival information system") for any organization that accepts "the responsibility to preserve information and make it available for a designated community." The focus of the standard is on the functions needed for long-term preservation. In the paper era, we tended to call such organizations "archives" and differentiated them from "libraries." The distinctions between traditional archives and traditional libraries and organizations that accept preservation responsibilities blur in the digital age. We address that blurring explicitly in Chapter 17 and expand on it further in Part Four.

OAIS is Actually Quite Simple!

While OAIS contains a lot of details, its strength is its essential simplicity. If you give any two or three experienced archivists 30 minutes to develop a functional model for long-term preservation, they will likely come up with the same basic concepts that are in OAIS. These concepts simply define the roles within the information lifecycle. OAIS calls this the "environment model."

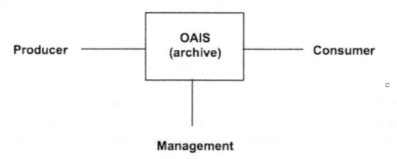

Graphic 15.1. The Environment Model of an OAIS

This environment model defines functional roles in the information life cycle as shown in Graphic 15.2.

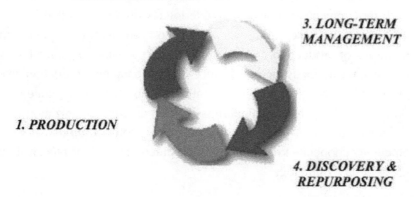

Graphic 15.2. The Information Lifecycle.

Preservation is not just the storing of information. An archive is needed to actively manage the preserved information and ensure that it can be found, understood, and used by the "consumers" of that information—its Designated Community.

This environment model may seem obvious, but unfortunately, it is very easy to ignore its implications in practice. For example, OAIS says that the roles of producer and archive are distinct and may even conflict with each other. The implication has direct application to the preservation of government information if we rely on agencies to preserve their own Public Information. Decisions driven by the "business needs" of an agency (the producer), the need to produce and disseminate Public Information quickly, and a focus on short-term access may conflict with decisions needed to create information that can be reliably preserved (managed) and used (by consumers) over the long term.

OAIS does not require the separation of the production of information and the archiving of that information into different organizations. But it does say that successful preservation requires an explicit preservation role. If we expect an agency that produces information to preserve its information and conform to OAIS, we would expect that agency to be legally required to do so and have an explicit preservation policy as part of its congressionally mandated mission. Few government agencies have such requirements or policies. Simple logic would suggest that it is less risky, more reliable, and even more cost efficient to separate the functions of production and preservation into separate organizations so that each can focus on its distinct role. Separating production from preservation also gives each organization the ability to concentrate on its primary (and legally mandated) mission. Two different organizations with different functions will be able to accurately communicate and efficiently cooperate to accomplish their different functions effectively by using OAIS terminology.

The Heart and Soul of Preservation: The Designated Community

The concept of the "Designated Community" is fundamental to OAIS. It is in every section of the standard and is mentioned over 75 times. OAIS is all about choosing a Designated Community and meeting the needs of that community. OAIS defines a Designated Community as "[a]n identified group of potential Consumers who should be able to understand a particular set of information" (OAIS section 1.7.2).

Essentially, OAIS says that the first question an archive must answer is "For whom am I doing this?" The answer to this question leads to other questions: What information will be selected for the archive? How will the archive ensure that the information is discoverable and understandable by the community?

A Designated Community can be very small and specialized or very large and general or anything in between. An archive can have more than one Designated Community, and a Designated Community can be composed of multiple user communities. An archive can even have another archive as a Designated Community.

The concept of preserving for a community of users is essential in the digital age. Unlike paper and ink books, which everyone used the same way, digital content comes in many forms and can be used in many ways. This is one of the inherent differences between digital information and printed information (see Chapter 16). There is no one-size-fits-all solution to content discovery, delivery, and use in the digital age. Different communities of users want to search for, acquire, and make use of digital content in different ways. Although the concept of one "general public" (44 USC 19) as the target community for "publications" was sufficient in the paper era, in the digital age the general public is only one of many potential target communities for published "Public Information."

OAIS Is for the Long Term

OAIS archives are concerned about the preservation of information "indefinitely" and "permanently." It defines the "Long Term" as "[a] period of time long enough for there to be concern about the impacts of changing technologies, including support for new media and data formats, and of a changing Designated Community, on the information being held in an OAIS. This period extends into the indefinite future" (1.7.2).

OAIS ranks the permanence of the information as more important than the permanence of the institution that preserves it. Both OAIS and TDR require that an archive have a succession plan (OAIS 3.3.6, TDR 3.1.2.1) to assure the permanence of the information even if the institution fails or changes its mission. This gives preserving institutions flexibility because an institution can design an OAIS archive to be intentionally temporary as long as it is part of a chain of custody that ensures the long-term preservation of information. There are, for example, "staging repositories" whose designated community is another archive that will take over long-term preservation of the content ingested and "staged" by the staging repository (Steinhart et al., 2009; Khan et al., 2011). One could easily imagine a government agency acting as a staging repository for the content it creates. Such an agency would ensure the preservation of its information while it is needed for agency business, but it would hand that content over to another preservation institution for permanent preservation when the content is no longer needed to conduct agency business.

(TDR has a slightly different focus than OAIS and includes a requirement that the institution must be sustainable [TDR section 3.4.1]. See Appendix A for a brief history of the development of TDR and its introduction of the criteria for sustainability.)

More Than Storing Bits

In order to preserve digital information for the long term, it is necessary to have some functional criteria to establish how we can know if the information will be usable in the future. For digital content, this means that OAIS is concerned with more than the storage of ones and zeros. It is concerned with guaranteeing access to and understandability of the bits preserved (2).

This means that in order to claim that information is being successfully preserved, we must be able to say that it is:

- Not just preserved, but discoverable. [2.2.2]
- Not just discoverable, but deliverable. [2.3.3]
- Not just deliverable as bits, but readable. [2.2.1]
- Not just readable, but understandable. [2.2.1]
- Not just understandable, but usable. [4.1.1.5]

As mentioned earlier, these guarantees are similar to the FAIR principles for scientific data management and stewardship that "all research objects should be Findable, Accessible, Interoperable and Reusable" (Wilkinson et al., 2016).

OAIS does not define how to make information usable, but it does provide the context for determining if it is. The key here is that the archive role is different from the producer and consumer roles. The role of the archive is one of addressing the functional needs of preservation of information and the use-needs of the community. This role is active, not passive. In the print era, preservation could be accomplished with little more than accepting what the producer gave and passing that along to users as originally produced. This was, in fact, a key element of the paper-era preservation infrastructure of traditional libraries and the FDLP. But in the digital age, preservation often requires more active transformation and repackaging of information to ensure its preservation and to meet the use-needs of a Designated Community of consumers now and new communities in the future.

Conforming to OAIS

Although OAIS does not prescribe solutions, it does specify two criteria that an archive must meet to conform to OAIS (1.4). An archive is required to:

- Conform to the OAIS Information Model
- Fulfill six OAIS Responsibilities

The Information Model = Content + Metadata

It takes almost 40 pages to completely describe the OAIS Information Model in detail [2.2 and 4.2], but in a nutshell, the model boils down to "Content plus metadata." This model will be familiar to librarians who are familiar with many kinds of metadata—particularly descriptive metadata and rights-management metadata—that point to information packaged as books and journals and journal articles and digital objects and so forth.

It is beyond the scope of this chapter to describe the OAIS Information Model in detail. Instead, we present here a quick overview of a few essential OAIS Information Model concepts focusing on content and metadata and how they are "packaged."

Content. OAIS describes "Content Information" as "the original target of preservation" and defines it as a data object (such as a PDF file) plus its associated "Representation Information" that turn the bits of the object into usable information.

Metadata. OAIS rarely uses the term "metadata." Instead, it describes different kinds of data (information) that an archive must preserve. Some of those kinds of information (descriptive information, rights information, administrative information) are often called metadata in other contexts. OAIS defines four broad categories of this kind of information:

- *Descriptive Information* describes the content being preserved with identifiers (such as titles, authors, subjects, etc.) and is typically used for both discovery and access [4.2.1.4.4].

- *Representation Information* is used to convert bit sequences (such as the raw ones and zeros of a jpeg file) into meaningful information (such as an image) [4.2.1.3]. This may be as simple as a statement that a particular bitstream (such as a jpeg file) conforms to a public standard (such as the jpeg standard, (International Telecommunication Union, 1992)).
- *Packaging Information* is the information that enables bits to be read from a specific medium (such as a disk, a tape, or a DVD) [4.2.1.4.3].
- *Preservation Description Information* (PDI). This includes reference, context, provenance, fixity, and access rights information [4.2.1.4.2] and is essential for the long-term management of the information in the archive.

Information Packages. The OAIS information model describes "packages" of information that include the content and metadata as described above. Content may need to be packaged differently for different functions (i.e., ingest and preservation and delivery) so OAIS describes a type of package for each function:

- *SIP. Submission Information Package*. The package that is sent (submitted) by a Producer to an OAIS for ingest into an archive. Its form and detailed content are typically negotiated between the Producer and the OAIS. Most SIPs will have, at minimum, Content Information and some PDI. [2.2.3]
- *AIP. Archival Information Package*. This is the package that the archive uses to store the information it ingests. This package need not be the same as the SIP, but it can be. Each SIP might produce a single AIP, but more complex combinations are possible. One or more SIPs might be transformed into one or more AIPs for preservation. For example, a SIP containing a "book" might be stored as several "chapter" AIPs, or several "chapter" SIPs might be combined into a single "book" AIP. Another example: all the elements of a web page (html text, CSS, JavaScript, images) harvested by web archiving software

might be stored in a single WARC "file," or each of its elements might be stored in separate WARC files, (US Library of Congress, 2020a) [2.2.3]. There is no single correct way to do this. Two different archives might choose different packaging strategies.

- *DIP. Dissemination Information Package.* In response to a user request, the OAIS provides information to a Consumer in the form of a DIP. The DIP need not be the same as the AIP or the SIP, but it can be. The DIP may be assembled from part of an AIP or from several AIPs. A DIP is analogous to what the Office of Management and Budget calls an Information Dissemination Product (IDP) (see Chapter 3).

There is a great deal of detail in OAIS about the Information Model, but even the few simple ideas described above should provide an insight into some of the everyday decisions that an archive must make.

For example, OAIS does not specify how an archive accepts, or stores, or delivers information packages—it only specifies that it fulfill those functions. OAIS specifies the function of these different packages. It makes the preservationist think about the functionality first and format or media or file-type only as a means to an end of acceptable functionality.

The Information Model focuses on functionality, not format. There is no "OAIS certified SIP format," for example. An archive does, however, need to specify what formats it will and will not accept from producers. These could be very specific standards-based formats, such as PDF/A (US Library of Congress, 2020d) or the Journal Article Tag Suite (US Library of Congress, 2020c), or very open-ended (such as anything harvestable from the web).

The Information Model does not require the archive to use different packaging and formats for the three functions (ingest, storage, and delivery of content), but it allows it to do so. The archive should let functionality determine the format of each package. Different archives might well make different decisions about how to package the same content. For example, two different archives might, for justifiable reasons, ingest, store, and deliver the same web page differently. OAIS just requires that the archive ensures that each package fulfills its function. The archive must ensure that the packaging guarantees that content will be discoverable and usable by its Designated Communities now and in the future.

The OAIS information model provides a common terminology that will help preservationists discuss digital library and digital archiving issues (e.g., selection, acquisition, preservation, discovery, service, delivery) with different communities. Information Technology professionals, database administrators, subject bibliographers, archivists, reference service providers, business office managers, and library and archive administrators do not all need to have a detailed understanding of digital formats or the OAIS Information Model. They should, however, understand the need to deliver preserved information in a way that matches the use-needs of their Designated Community. And subject specialists in vastly different areas (e.g., history, political science, agriculture, economics) can speak with each other clearly about the needs of their different Designated Communities and describe those needs—however different they may be—using the same terminology for ingest, preservation, and dissemination.

Six Responsibilities

The second mandatory OAIS requirement is that the archive must fulfill six responsibilities (3.1).

1. Negotiates for and Accepts Information
2. Obtains Sufficient Control for Preservation
3. Determines Designated Community
4. Ensures Information is Independently Understandable
5. Follows Established Preservation Policies and Procedures
6. Makes the Information Available

These responsibilities are essential to a complete lifecycle of information (Graphic 15.2). All of the six responsibilities must be addressed adequately, or information will be lost and the lifecycle will be broken.

The Functional Model

To help archives design systems that adequately meet the mandatory requirements, OAIS provides a Functional Model to complement the Information Model.

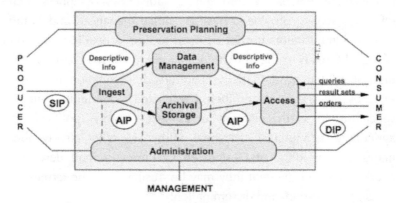

Graphic 15.3. The "Functional Entities" of Archival Management

The shaded bubbles in Graphic 15.3 are the six functions of an OAIS compliant archive:

1. Preservation Planning
2. Ingest
3. Data Management
4. Archival Storage
5. Administration
6. Access

OAIS describes the Functional Model in some detail, but the concept is simple. This functional model is consistent with the Environment Model and the Information Model, and are all reflected in the three stages, the three packages, the processes, and the functions.

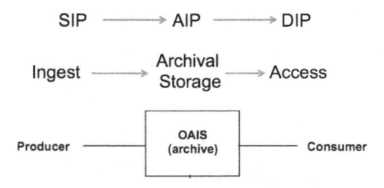

Graphic 15.4. The three parts of preservation

The Functional Model in Graphic 15.3 provides details of the management functions of the archive. The six "functional entities" of management (the shaded bubbles in the figure) are related to each other within the archive and relate externally to the producer and consumer of information. This functional model shows the full scope of the management function of the archive that was abbreviated in the Environment Model (Graphic 15.1). These management functions are essential to long-term preservation.

Summary

OAIS provides clear definitions of the functions of long-term preservation by focusing on the information and its use by user communities. By focusing on the functions of preservation instead of the traditional roles of existing organizations, it opens up opportunities for organizations that are ready to adapt to the digital age. This is particularly important for government information, which includes traditional publications that are the Records of government and Records of government that, in the digital age, function more and more like traditional publications. OAIS gives us new terminology that makes it easier to understand how accessibility for different communities is an essential component of preservation.

CHAPTER 16
Characteristics of Digital Information

Inherent Characteristics of Digital Information

In previous chapters, we have referred to the inherent characteristics of print and digital information and suggested that those characteristics affect their preservation. In this chapter, we define those inherent characteristics of digital information that are different from those of print information and describe their implications for preservation.

Did Digital "Change Everything"?

The fact that digitally published information is characteristically different from printed information is so well known as to be a truism. Indeed, since at least the early 2000s, the FDLP community has largely accepted a trope that digital information (or, sometimes, the internet, or the web, or just technology in general) has "changed everything." (See for example: a Superintendent of Documents who called this "the ICE Age—Internet Changes Everything" [Buckley, 2000a]; the Public Printer in addressing funding and business models who said, "Everything has changed" [James, 2003]; another Superintendent of Documents who described "a paradigm shift in preservation of depository materials" [Davis, 2008]; and the 2011 Ithaka S+R study of the depository system that cited a "paradigm shift" that was causing "upheaval" in the depository community and that would cause a shift from a "production economy" to a "service economy" [Housewright & Schonfeld, 2011; Arrigo, 2004]).

These expressions of inevitability were not so much predictions of the future as they were retrospective rationales for specific policy changes

197

already made by GPO in 2000 and, as such, they were close to what Robert Jay Lifton calls a "thought-terminating cliché," whose purpose is to dismiss dissent (Lifton, 1961). That policy change was intended to provide "access" to electronic information, but it did so by effectively ceasing deposit of government information with depository libraries (Buckley, 2000b). This was not a trivial policy adjustment but a 180-degree reversal of 200 years of policy and practice. As recently as 1972 GPO had demonstrated that it had no role at all in preservation when it gave its entire Public Documents Library (dating back to 1895) of almost two million volumes and 76,000 maps to NARA (Barnum, 2011). (As noted in Chapter 5, access to this collection "of last resort" is limited.) But with this new policy, GPO took sole responsibility for electronic information access and "permanent access," if not preservation (Jacobs, 2000). (See Appendix B for details of this policy shift.) Under the new policy, depository libraries no longer had a role in preserving government information.

This reversal in policy was largely accepted by FDLP libraries at the time. Indeed, when GPO developed its online storage facility (called the "Federal Digital System" or FDsys), the Association of Research Libraries (ARL) simply accepted the reversal, saying that "the roles of GPO and participating libraries fundamentally changed" (Association of Research Libraries, 2009). In the late 1990s and early 2000s, few libraries had the digital tools, staffing, or experience to select, acquire, organize, store, and preserve digital information or to provide digital services for that information. So, when GPO said that it would do all of that, most FDLP libraries accepted it without question and, most likely, with some relief. It was no doubt comforting to the managers of library budgets to say that technology had "changed everything," that technology made this particular policy inevitable, and that it would not cost the library anything.

But this vague and ill-defined excuse was incomplete and misleading, and it led to the acceptance—and justification as inevitable—of policies that have failed to adequately preserve government information (see

Chapter 13). Saying that the digital shift (or something) had changed everything implied that technology dictated the response to those changes. As GPO's electronic collection manager said, "technology" meant that GPO "must" do this:

> In terms of assuring permanent access, technology has radically altered the deposit model: where previously permanence was assured by multiple depositories being bound to retain publications in perpetuity, and GPO's responsibility as custodian was largely complete at the point that publications were shipped to depository libraries, GPO must now assume responsibility for keeping the single electronic source not only currently available, but technologically viable. (Barnum & Kerchoff, 2000)

While the digital shift was real and disrupted the paper-era infrastructure and required a new infrastructure, it did not dictate these specific policies. We strongly disagree with any presumption that technologies dictate their own use. As historian Yuval Noah Harari says, "Belief in technological determinism is dangerous because it excuses people of all responsibility" (Harari, 2024). Indeed, since the preservation gaps have widened under those new policies, the policies and their inevitability must be revisited. It is certainly true that the digital shift required the creation of a new digital preservation infrastructure, but the policies chosen failed to do that.

Of course, a lot of things did change with the introduction of digital publishing and the wide availability of the web, but a lot remained the same. Although digital publishing has the potential for revolutionizing how we record and transmit and use information, a lot of digital publishing is explicitly intended only to emulate printed documents. John E. Warnock, the co-founder of Adobe, described the Adobe PDF format (which is the second most common format on dot-gov webspace) as being designed to make page-oriented documents that could be viewed and printed (Warnock, 1995). What we need today is an understanding of what actually did change, what effects—both positive and negative—those changes have on preservation, and what are the available responses to those changes that can minimize the negative effects and maximize the benefits.

Challenges

Three of the inherent characteristics of digital information are particularly well known and are often discussed in relation to the challenges they present to preservation. These are that digital objects are malleable, fragile, and subject to obsolescence.

- Malleable. Digital objects are easy to alter. Indeed, a lot of digital information is intentionally dynamic, changing over time or use. This malleability extends to making digital content easy to delete entirely.
- Fragile. Common digital storage media are not permanent, and digital objects can become corrupt because of "bit rot."
- Subject to obsolescence. Digital objects can become difficult or impossible to read because of technological obsolescence. Being able to use a digital file in the future is dependent on rapidly changing, interlocking technologies including hardware, operating systems, storage media, file formats, and software.

As is well known, these digital characteristics present challenges to preservation that do not in general exist for print. Of particular concern is malleability. Digital objects may be intentionally updated, but they can also be unintentionally altered or damaged beyond repair or intentionally altered or even forged. To ensure accurate preservation, new practices and standards are necessary at all three archival stages (ingest, maintenance, and delivery). Typically, digital objects also require more metadata than print required. This includes, most notably, the four categories of metadata that OAIS calls descriptive, representation, packaging, and preservation description information (see Chapter 15). Digital preservation also requires checking for bit rot by doing integrity checks of digital objects using fixity information. To deal with obsolescence, preservationists typically adopt strategies such as migration of file formats and emulation of software and operating environments.

All of these procedures and practices require either new infrastructures (procedures, practices, standards, etc.) or additions to and modification of the existing infrastructure. By coping with these three inherent differences, preservationists can ensure that digital objects can be successfully stored without unintended alterations and that those objects will be, as Barnum and Kerchoff said, "technologically viable" in the future.

Addressing the Challenges

In 1993, when the *GPO Access Act* (Public Law 103-40) directed GPO to create an "electronic storage facility," GPO began a long process of creating a Trusted Digital Repository that would address these inherent challenges of preserving digital information. It not only acquired the hardware and software for the repository, but it also developed policies and procedures for an internal infrastructure for preserving digital content and for making it publicly available. The resulting storage infrastructure (GOVINFO) earned ISO 16363 Trusted Digital Repository certification in 2018. GPO's development and operation of GOVINFO is a significant accomplishment, and it does an admirable job of addressing the three preservation challenges listed above.

GOVINFO is arguably the most significant step to date toward the development of a preservation infrastructure for government information, but a limited one. A complete DPI will require more than a single trusted repository. It will require laws, regulations, accepted norms, organizational arrangements, and standard practices—the essential elements of an infrastructure as described by Star and Ruhleder (Star & Ruhleder, 1996). And it will require many partners working toward a common goal. It would be unreasonable to expect any one agency to be able to build a preservation infrastructure for all government information on its own.

The limits of what GPO can do are apparent in its coverage and scope. As described in Part Two, GOVINFO lacks most of the information published by the executive branch and GPO's web harvesting currently includes only about 1 percent of government websites. As GPO itself has said, it cannot keep pace with the scale of agency publishing directly to the web (GPO, 2017).

But perhaps even more significantly, GPO's digital services for the information it does preserve are, by design, very limited. Its definition of its Designated Community does not include the general public (PTAB Primary Trustworthy Digital Repository Authorisation Body Ltd, 2018). It relies instead on a community of depository librarians to intermediate between the public and govinfo.gov (GPO, 1996b; Arrigo, 2004). But it gives those librarians and their libraries no way to build their own collections or services, and its discovery and access tools are limited.

In the next section, we describe two other inherent characteristics of digital information that provide opportunities for addressing such limitations.

Opportunities

There are two other inherent differences between paper and digital information that offer not barriers but opportunities for enhancing preservation: utility and packaging. Although these differences are well-known and understood, they are not yet integrated into a digital preservation infrastructure. In spite of their familiarity, it is necessary to review them in order to describe their implications.

Utility

From the most basic functions, such as searching digital text and the ability to copy and paste, even the simplest digital object has properties lacking in print publications. In comparison, the functionality of print publications is very limited.

Print publications have a single utility: they can be used (text read and graphics viewed) by looking at them. Digital publications, on the other hand, can be used in many different ways. Foremost in new utility is the availability of different file formats that are designed for specific uses: PDF for reading documents, GIS for creating maps, spreadsheets for using numbers, datafiles for statistical analysis, JSON for information exchange, and so forth. These formats are designed to convey information that can be summarized, analyzed, visualized, and combined with other information to create new information in ways that were difficult, inconvenient or impossible for print-based information. When one considers that web harvests have identified as many as 500 different MIME filetypes on government web servers (Chapter 6), the variety of uses is limited only by the imagination.

In addition, digital formats can be used in ways that may not have been anticipated by the agency that produces them. Different users can make use of the same content in different ways. One user may want to use GIS data to simply look at a static map, while another user may want to combine geolocations with population data to make a new map. One user may read a statistical table to get a single number, while another would prefer to create new information by analyzing or summarizing or visualizing or sorting the statistics. Even static digital documents can be used in more ways than printed documents. One user may wish to simply read a document, much the way a printed document was read, but a different user may want to computationally analyze the text in that same document. It is an inherent characteristic of digital information that it has the potential to be used, reused, and repurposed in many different ways.

Packaging

The way digital information is "packaged" is a second characteristic that also affects its utility. Printed publications are packaged in a single physical medium that is used across the lifespan of that information for storage, transmission, and use. One simply uses the single physical medium of "the book" (or its other equivalent formats such as pamphlets, maps, etc.) across the lifespan of the content. In contrast, the storage, transmission, and use of digital publications are not tied to any particular medium or format or bundle of information. Information can also be packaged in different ways. For example, a single set of census data can be published as statistical tables or data files for analysis, or graphs, or maps, and so forth. Information can also be grouped into different packages (e.g., "files") for different purposes and files can be split into more granular units of content or combined with other units in useful ways. Content in the digital age need not be confined to a single "package" such as a book or a periodical; it can be split into smaller packages or combined into specialized packages. For example, users need not be limited to getting a digital package of an entire published issue of the *Federal Register* in order to read a single announcement or proposal. They should be able to retrieve a package of content that has just what they need, whether it is a single notice from one issue, or a series of related notices, proposals, comments, revisions, and final postings from many issues that were published over a period of years.

An individual package of digital content can also be combined with other packages to repurpose or recontextualize the information. While printed books must be used one at a time, collections of digital files may be combined for computational analysis or to create new information. One user may wish to use a single GIS file to view a contemporary map of a county, while another may wish to combine several GIS files in order to create an interactive view of population movements within that county over time. One user may wish to use census data to look at the population of one county in a particular year, while another user may wish to combine those data with water and air quality data and federal infrastructure expenditures in order to analyze the effectiveness of governmental policies. Such combinations are not limited to information from a single agency or "publication." Compatible data sources can be found across publications and agencies and from non-government sources.

For printed publications, such repackaging and repurposing of content is normally only done by publishers and done in very limited ways. Individual Acts of Congress are republished in volumes of the *Statutes at Large* and then rearranged and codified and republished as volumes of the *United States Code*. House and Senate Reports and Documents are published individually and then combined into the *Serial Set*. The same census data that was used to create bound volumes of statistical tables for counties was used to create other bound volumes of statistical tables for block groups. Individual users cannot in general repurpose or recombine printed publications, but such repackaging and repurposing of information by individual end-users is common and an inherent feature of digital content.

OAIS recognizes this by defining different packaging of information for ingest (SIP), archival storage (AIP), and delivery (DIP). Although these packages can all be the same (as they are for printed content where "the book" is the package for all three functions), they can also be different for each function. This empowers the archive to ingest information in the most efficient way, store information to maximize securing the content for the long term, and deliver content in packages that match each use-case need for members of different Designated Communities.

Implications

The inherent characteristics of digital information that differentiate it from printed information have implications for users, for libraries and archives, and for a future digital preservation infrastructure.

The malleability, fragility, and risk of obsolescence of digital information have already led libraries and archives to adopt new practices and procedures to ensure the integrity and longevity of the information they preserve. The print preservation infrastructure did not have to deal with these inherent characteristics. As archives and libraries address these characteristics, they are beginning to develop pieces of a digital preservation infrastructure.

Preservationists now need similar infrastructure-scale practices to take advantage of the flexible utility and packaging of digital information for the benefit of users. These inherent characteristics affect how users will want to search and browse for information and how they will want it packaged for retrieval and use. A digital preservation infrastructure must support users who want to use information, not just view it as published.

Print-era solutions to searching and browsing for information and for identifying and delivering information are simply inadequate for digital information. In the print era, information was defined by its physical instantiation as a book. "The book" *was* "the information." There was only one use-case of the information: viewing the printed object. Users could search and browse for information in simple ways because the target of their discovery and the information that would be delivered was a physical object defined by its publisher.

But the publisher's packaging of information is no longer the sole definition of the information. Information is defined by metadata and, in a digital preservation infrastructure, will be definable by users. It can be delivered in different packages and used in different ways. If being "malleable" is a challenge, then the flip side of that coin is that digital information is also flexible. This means that libraries and archives will need to provide discovery tools that turn discovery around. Instead of forcing users to look for objects defined by a publisher, they need to provide discovery tools that allow users to define information packages that meet their use-needs.

As Paul Conway said, "In the digital world, the concept of access is transformed from a convenient byproduct of the preservation process to its central motif" (Conway, 1996). Pointing at static objects is not sufficient "access" for digital information. Providing flexible accessibility to information is not optional; it is an integral component of digital preservation. This is why the definition of preservation that we use in this book explicitly includes discovery, delivery, and usability.

> Preservation means maintaining information for discovery,
> delivery, readability, understandability and usability by a
> Designated Community for the long-term. (Chapter 4)

A preservation infrastructure that addresses new needs of information users must also address the selection needs of libraries and archives. Libraries and archives need an infrastructure that empowers preservationists to select information for preservation based on how its utility and packaging match the needs of its Designated Communities. OAIS recognizes this, saying that the archive "should negotiate with the Producer to ensure it acquires appropriate Content Information and associated PDI for its mission and the Designated Community."

In the print era, the publisher defined what was available for preservation. In the digital age, the target of preservation is no longer necessarily and only a single, viewable presentation and packaging provided by the publisher, but the usable information itself. Libraries and archives need a preservation infrastructure that enables them to identify the packaging of information that meets two criteria: it should be easy to preserve, and it should enable the library or archive to meet the use-needs of their Designated Communities now and in the future.

In the next chapter, we review old assumptions about a preservation infrastructure in the context of these implications.

CHAPTER 17
Assumptions

It is clear that the different inherent characteristics of printed and digital information require different preservation strategies, and a successful preservation infrastructure needs to accommodate those strategies. It is less clear what the elements of a new infrastructure are or how the new infrastructure will affect libraries and archives. Before we start envisioning those elements, it is worth reviewing some of the print-era terminology that preservationists still use and the assumptions behind that use. The connotations of terms that got their meaning in the print era do not always fit in the digital age. The way we use language and the assumptions we bring to that use affect how we think about the world. How we envision the future should affect how we use language as well. In this chapter, we briefly address a few of the most significant old assumptions and terms.

- Communities and Constituents
- Archives and Libraries
- Availability and Accessibility
- Information and Information Service
- Collections and Services
- Addresses and Identifiers

Communities and Constituents

In the print era, the community that a library served had to be geographically close to the library's physical facility. That led communities such as universities and local governments to build brick-and-mortar libraries that served their communities. The physical limitations of printed collections led logically to libraries that supported

208

a geographically based community defined by a financial and administrative constituency. This is a one-to-one model: each one local library supports only one community, and each community is supported by its one local library. In the print era, librarians could assume this was optimal.

In the digital age, when access no longer requires geographic proximity, libraries have the technical flexibility to define their communities differently. Although licensing agreements and copyright restrictions on commercially published information often constrain who a library can serve, those commercial constraints do not exist for government information since it is mostly in the public domain. In fact, FDLP libraries are required by law (44 USC 1911) to make their deposited government information available to the general public even when other collections are restricted to the library's constituents. Together, technology and the law give libraries the flexibility to provide government information collections and services to communities beyond their local constituency.

Libraries in the digital age can choose to serve communities based on the shared interests of those communities. They can tailor their information services to match the ways their communities search for and use information. The individuals that comprise such communities will no longer have to be geographically close to each other or to a brick-and-mortar library. Libraries will, of course, still be obligated to serve their financial and administrative constituencies. Libraries will continue to do that but will have more flexibility in how they do it. For government information, libraries are no longer locked in to a one-to-one service where each library must be the sole provider of all information services to one local community. Many libraries can serve many communities. Each library can provide services to more than one community, and those communities can be anywhere. Each local constituency can receive services from many libraries. This many-to-many model will reach more communities and increase services available to all communities.

A new DPI will need to support sharing and interoperability of collections and services, natively.

Archives and Libraries

In the print era, our assumptions about the roles of libraries and archives were clear and distinct, even for government information. Traditionally, libraries and archives have played different roles in the information ecosystem. Government Records and Government Publications were preserved differently and made accessible differently. Librarians and archivists were trained differently. But in the digital age, the distinction between unpublished Records and published Public Information and between the roles of libraries and archives are blurring. This presents both challenges and opportunities for preservation of government information. Our old assumptions of the distinct roles of libraries and archives do not work as well as they did in the paper era.

The tasks of libraries and archives are quite similar. They select, acquire, organize, and preserve information and provide information services to their communities. But they tend to approach these tasks differently and tend to serve different communities.

Archives tend to be tied to records management and libraries to publishers. Archives tend to have primary materials; libraries tend to have secondary and tertiary publications. Archives tend to aim for completeness; libraries are more selective. Archives acquire authentic records regardless of their accuracy; libraries select information that they judge to be accurate and current. Archives tend to organize collections by provenance and tend to focus on the context of the creation of the record; libraries organize their collections by subject and focus on the information content. Archives treat their materials as unique or the copy of last resort; libraries treat their materials as consumable and replaceable and even potentially discardable. Archives tend to provide discovery tools that focus on broad collections and groups of documents; libraries focus on individual titles. Archives prioritize protecting collections from damage or loss, which tends to limit access and use of the materials and which, in turn, tends to limit the communities they serve; libraries tend to have open stacks and serve a broad public. Archives tend to serve communities that are interested in information because of its source and original purpose; libraries tend to serve communities that are interested in the information value not necessarily related to the origin of the information or its original purpose (Borgman, 2007a). Some unpublished Records have legal restrictions on access in archives; published Public Information has no such restrictions in libraries.

Peter Hirtle, Technology Strategist at Cornell University and former coordinator of electronic public access at NARA, summed up the differences between libraries and archives this way: "Archivists deal not in information, but in records," and "Government records are different from government information" (Hirtle, 2001).

But, even in the print era, these seemingly clear differences were less clear for government information. Government Publications deposited with FDLP libraries possess many of the characteristics of government Records and some government Records possess characteristics of publications. Rodney Ross of the National Archives described this overlap, saying that Records are byproducts of business activity, that library materials are "cultural products," and that Government Publications are both (Holterhoff, 1990a). This overlap becomes explicit when an agency deposits its government Records with NARA; those Records without access restrictions become, by definition, Public Information. The similarities of government Records and Public Information also led FDLP libraries to treat publications more like Records. FDLP libraries selected and acquired publications by agency and publication type rather than by subject (FDLP, 2023a). They had no mechanism for selecting publications on subjects such as "water quality" or disciplines such as economics, or geographies like "the southwest." And many depository libraries separated their government documents into a separate department and organized them by provenance (that is, by agency, using the Superintendent of Documents classification [SuDoc] system) (Richardson Jr et al., 1980). FDLP libraries may not weed their Government Publications without going through a careful procedure—including alerting their regional library of titles being considered for withdrawal—that was meant to ensure that the information would not be lost and that it would remain publicly available (GPO, 2018a). Libraries serve communities that use Publications both for their information content (following the library model) and as evidence of the activities of agencies (following the archives model).

In the digital age, these similarities increase and the differences between libraries and archives diminish. This was a recognized issue as early as 1988 when the Office of Technology Assessment issued its well-known report "Informing the Nation: federal information dissemination in an electronic age." That report said that "[a]t a fundamental level, electronic technology is changing or even eliminating many distinctions between reports, publications, databases, records, and the like, in ways not anticipated by existing statutes and policies" (Office of Technology Assessment, 1988). Hirtle qualified his distinction between "information" and "records" by suggesting that "digital libraries will need to adopt some of the principles of archives" and "the lines between libraries and the traditional repository for government records (archives) are blurring." He concluded: "[G]overnment archival records can be integrated into the emerging digital library in ways that have never been considered in the past."

Archives have adapted their procedures to the digital age, changing how they provide services and even who they serve. Archives use digital accessibility tools, including MARC and the Encoded Archival Description standard for archival finding aids. People who, in the paper era, had to visit archives to use their collections can now look for, find, and consult digital originals and digital surrogates of archival records online. People who in the past never used archival materials and may not have even been aware of them have become frequent users of the digital collections of archives (Martin, 2003). NARA has committed to digitizing all of its analog holdings, making them available to the public online, and making it easier "for members of the public" to search and discover relevant records (NARA, 2022a). NARA is changing its more passive role "of making records available for others to discover" and taking a more active role of "making access happen by delivering increasing volumes of electronic records to the American public online, using flexible tools and accessible resources that promote public participation" (NARA, 2021d).

These changes make many of the paper-era assumptions about the distinct roles of libraries and archives inadequate.

The digital age has brought two specific changes that will affect not so much what libraries and archives do but how they do it.

The first of those changes is the shift to digital publishing, which changed the inherent characteristics both of unpublished Records and of published Public Information. Both can now be easily copied, but both can now also be easily altered. It has become technically easier to make more information of both kinds available for public inspection. But the characteristics and the sheer volume of information produced make it more difficult to manage and preserve Records and Publications. This means that libraries and archives need to manage information as opposed to individual printed documents. Information management includes procedures for identifying information and tools for discovery and delivery of information. Libraries need to provide archival context and provenance for Public Information—even if they are providing access to a small subset of all Public Information. Libraries and archives need to collaborate to ensure that a complete record of published Public Information is preserved, not just the 1 percent of unpublished Records that was traditionally preserved by NARA. Archives need to go beyond the needs of those seeking static Records and anticipate the discovery and use-needs of new communities of users and facilitate new digital services that meet those needs. Both need to ensure that preserved information can be discovered, delivered, and used in a variety of ways for a variety of communities.

The second big change is that Congress explicitly directed NARA to treat published Public Information as Records. This results in the problems of applying to published Public Information the rules for retention and the mechanisms of access designed for unpublished Records as described in Chapter 5.

As the information that libraries and archives manage continues to merge and as the methods that libraries and archives use to provide services overlap, our old assumptions about the roles of libraries and archives need to evolve to accommodate these changes.

A digital preservation infrastructure will need new assumptions in order to provide both the essential functions of archives (ensuring the integrity, completeness, authenticity, and usability of Records) and the essential functions of libraries (making the information content of Public Information accessible and usable to many different communities). As the content managed by libraries and archives increasingly overlaps and as libraries and archives increasingly offer similar digital services, the need for a new digital preservation infrastructure that supports both libraries and archives becomes essential. A successful DPI will not change libraries into archives or archives into libraries. It will instead enrich libraries with archival holdings and enrich archives with library services.

A new DPI will need to support libraries and archives that share content and manage complementary services.

Availability and Accessibility

Making information available is not the same as making it accessible. "Availability" is binary: something either is or is not available to the public. Accessibility, on the other hand, is not an either/or condition but can vary by degrees across several dimensions. A district court opinion available behind the difficult-to-use paywalled PACER system and a presidential proclamation on the White House website are both Public Information, and both are publicly "available," but they are not equally "accessible."

In the context of Preservation, it is necessary to distinguish between availability and accessibility. They are different but interdependent. Specifically, availability is the foundation on which accessibility is built. If a digital object is not available it can never be accessible, but making any given document available amidst the availability of hundreds of millions of other documents does not make it accessible. Once a digital object is reliably and permanently available, however, preservationists can develop all kinds of functions that make the object more accessible.

This distinction was true in the paper era, but its application to preservation is critical in the digital age. Accessibility can range from difficult to easy, from expensive to free, from just "technologically viable" (Barnum & Kerchoff, 2000) to usable in new and innovative ways. Information can be "available" online, but not be easy to find. As NARA has noted, "[S]imply posting records on a public website does not make them easy to discover or use" (NARA, 2021d).

One common use of the term "accessibility" refers to making information more usable by people with disabilities (Digital.gov, 2019). This usage is so common that, in some contexts, "accessibility" is assumed to mean *only* "accessible to a wider range of people with disabilities" (World Wide Web Consortium, 2008). This is one essential dimension of accessibility, and it provides a model for meeting the needs of other communities that need to use digital information in ways other than viewing. The distinction between availability and accessibility has more dimensions in the digital age because of the inherent characteristics of digital information that create new opportunities for searching, acquiring, and using information. This creates new communities of users with new use-needs.

The most significant of these other dimensions of accessibility are: offline/online, discovery, identity, delivery, usability, and cost.

The most commonly recognized of these is the offline/online dimension. Recognizing that accessibility is not binary, GPO has defined "accessibility" as a matter of "degrees" (GPO Library Programs Service, 1998) and says that it digitizes paper publications in order "to expand free public access" (GPO, 2021f). The paper editions were always available and accessible at libraries, but digitizing them "expands" accessibility by making them available online.

This offline/online dimension is so important that online availability is often mistaken for adequate accessibility. For example, GPO's policy that allows Regional Depositories to discard paper documents uses the term "access" 13 times. But, using the distinctions we describe here, "providing access" really means "make available," and "online access" really means "online availability." GPO even defines "permanent public access" as being "available for use" (GPO Superintendent Of Documents, 2020). The policy, however, makes no mention of enhancing accessibility in any other dimension, nor does it acknowledge that print and digital information have different accessibility features. When developing a digital preservation infrastructure, preservationists should not assume that online availability without enhanced accessibility is either complete or sufficient. Neither should digitization efforts assume that a digital surrogate necessarily replaces the usability of the original printed work.

While the offline/online dimension is very important, it is only the lowest bar of accessibility in the digital age. Libraries and archives must also ensure that users can find the information they need, not just the packages that contain it, and identify it and its relationship to other preserved information. They must be able to package and deliver information the way users want it and ensure that the information is usable.

In the print era, libraries and archives could assume that people would adapt to publishing and library and archival standards for discovery, identification, delivery, and use of information. In the digital age, users expect libraries and archives to adapt to the use-needs of the communities they serve.

Finally, the cost dimension of accessibility cannot be overlooked. Although most federal government information is available without a fee as we write this, cost has always been and still is a crucial dimension of accessibility. Most digital judicial information, for example, is available, but access is locked behind the paywall of the PACER system. There are other fee-based and restricted-access government information services as well (FDLP, 2022b; NTIS, 2022). The law that resulted in the creation of the govinfo.gov website allows GPO to charge fees for its use (44 USC 4102). Some information that is not available through the depository program or govinfo.gov is available through the GPO's sales program. There have always been (and still are) those who advocate limiting access to government information by imposing fees, privatizing GPO, or relying on the private sector for access (Jacobs, 2011). While the private sector can certainly add value to government information, its services must be understood as complementary to full and complete Preservation and availability of government information and not a substitute for it. The existing infrastructure does not guarantee complete, long-term, free availability of preserved information. In the absence of a digital preservation infrastructure that can guarantee free availability, all government information is at the mercy of relatively small changes in government policies and budgets.

Digital accessibility can best be understood from the perspective of users of information. A user will judge the degree of accessibility by asking the question, "I think that the information I want is available, but can I find, get, and use it?" This will lead the preservationist to measure accessibility by asking, "Am I providing adequate discovery, delivery, and usability for my Designated Communities?" This will affect assumptions about the success of preservation policies. Comforting statistics about numbers of bytes or files stored, numbers of users, and numbers of downloads will no longer be adequate for assessing the success of preservation. Archives and libraries will also need to find ways of measuring unmet needs and gaps in services and will have to find ways of identifying potential users.

A new DPI will have to do more than make static digital objects available. It must support new kinds of discovery, identification, acquisition, and use of information itself by new kinds of user communities.

Information and Information Service

In the print era, it was easy to equate "the library" or "the archive" with the information in its collections. In the digital age, preservationists must distinguish between information and information services. An information service is a gateway to information; it is not the information itself. The most familiar example that illustrates the difference is a bibliographic database. The database contains information; the user-interface to the database is an information service that provides access to the information. Theoretically there could be any number of user interfaces to the same information, each having a different look and feel and functionality, but each providing access to the same information. The information is, or should be, the primary target for preservation because the information service (the user interface) is not, itself, information. (Although there are legitimate reasons for trying to also preserve the look and feel of a particular service, doing so is a different task with a different purpose than preserving the information. See Appendix C.)

To be clear: when we write here of government information services, we are *not* referring to those "e-government services," which we view as those websites that provide *transactional services* to individuals and businesses. E-government services provide a way for individuals and businesses to transact business with the government (e.g., filing your tax return) using two-way communication. Such e-government transaction services are completely different from the one-way communication of information services that provide users with access to Public Information (such as looking up the population of a city on a Census Bureau website).

The distinction between information and information services is of practical importance to preservationists that use web harvesting to preserve Public Information. Websites are information services. Web harvesting attempts (to some extent) to preserve the information service as well as the information provided by the service. A key challenge for web harvesting is determining if the information service (the website) provides a mechanism for reliably harvesting all the information that is available. One of the reasons that NARA gave for stopping its own harvests of executive branch websites was that, while harvests "may provide some indication of 'look and feel'" of a website, "the web snapshot does not systematically or completely document agency actions or functions in a meaningful way" (NARA, 2008c). Preservationists must not ignore the fact that government information services are not all equally good (or bad). Duplicating existing good information services is unnecessary, and preserving bad user interfaces is counter-productive.

In Chapter 6, we mentioned an example of how web harvesting can fail to extract the information from the information service. The 2020 End of Term crawl captured the "look and feel" (Bower et al., 2012) of GPO's *Catalog of Government Publications* (CGP) website (catalog.gpo.gov), but it only captured 3 percent of the information available through that service. (Interestingly, an alternative option for preserving all the information does exist. GPO makes the entire CGP database available for download on GitHub [GPO, 2023a]).

A new DPI must enable libraries and archives to select and acquire Public Information precisely and comprehensively so that they can provide their own information services for discovery, acquisition, and use of that selected information to their Designated Communities.

Collections and Services

In the print era, libraries were identified with their collections. The scope of a library's collection defined it by type (e.g., public, school, undergraduate, law, medical, special, etc.). The size, breadth, and depth of a collection made each library uniquely valuable to its constituency.

Libraries were not *just* their collections, however. They were also the information services they provided to their communities. Service began with collection development, which consciously and intentionally built a collection designed for use by a community of users. It cataloged, arranged, stored, and maintained the collection. Human public services helped users find the information they needed in the collection—and beyond. Physical facilities helped people use the information they gathered. Collections and services were inextricably connected, and together, they defined an institution as a library.

Significantly, there was no such thing as a library without a collection. But in the late 20th century, a number of factors came together that led libraries to reexamine their assumptions about the composition and importance of collections.

By that time, many of the existing buildings that housed libraries' collections were full or nearly so. At the same time, there were increasing numbers of publications that fit existing collection development profiles, and their costs were rising faster than inflation. Library budgets were not growing fast enough to cope with increased costs of building collections, and libraries were literally running out of space to house more physical volumes.

Research that used data about collections began to demonstrate that there was a lot of overlap among library collections. Some books and journals were in many libraries, and librarians began questioning the need for that duplication. At the same time, scientific journals, which are typically the most expensive literature for academic libraries, were moving online and the model of owning a copy of every issue of every journal seemed, to many librarians, both redundant and unnecessary. As users began to prefer online access to physical access, librarians began questioning the value of building local collections of physical books and journals. The question librarians began asking was if those physical collections were needed at all, particularly when they were, apparently, not being used (Malpas, 2011).

Researchers began studying the feasibility of shared collections; instead of 100 libraries having 100 copies of the same books and journals, would it be more cost-effective to have three, or maybe five, copies, collectively managed? This would solve the library space problem, save collection funds, and allow libraries to provide access to the same or even more information. Ensuring that a few copies were preserved in a cooperative way would ensure the preservation of the information. Studies began comparing the costs of owning a collection versus the cost of borrowing items or purchasing them on demand (Ferguson & Kehoe, 1993). Paying for access to someone else's copy might be slower, but it was "more cost-effective than collection building" (Butler, 1987). Providing "just in time" access began to seem more cost effective than building a collection "just in case" someone wanted something that a librarian had anticipated them wanting (Schonfeld & Housewright, 2009b).

In this emerging model, preservation of information remained constant but was reconfigured as a shared responsibility across the library community rather than the responsibility of each individual library.

This led, in turn, to a new way of thinking about library collection development. Special collections—information that other libraries did not have—became "assets" that made a library unique and valuable. But there was less incentive to add to a collection those books that were easily available and already in collections of other libraries—or even in bookstores (Anderson, 2013). Some libraries started using a policy called "patron-driven acquisitions" (or "demand-driven acquisitions"). With this approach, a library would purchase materials only when one or more patrons asked for them (Anderson, 2011c). By shifting a large part of selection responsibility from librarians (whose perspective is the long-term collection-as-a-whole), to individuals (whose perspective is the short-term, immediate need of a single title), these policies change the very nature of libraries. Instead of the library being identified with its collection, the library was becoming a service without a planned collection.

Along with these developments came terminology (e.g., "book warehouses" and "book museums") that implicitly deprecated the value of physical collections at all (Anderson, 2011b). The old assumption that a library was its collection started changing to discussions of "the library as space" for teaching, research, collaboration, study, and community (Bennett & Council on Library and Information Resources, 2005). The rhetoric began to change and libraries began to describe a shift from being "collection-centric" to being "user-centric" (Giesecke, 2011).

The assumptions about the role of librarians and libraries began to change along with these developments. New assumptions began to arise that suggested that librarians would "evolve" from "a collections-oriented role to a service-oriented role." Government information librarians in particular would no longer curate collections but would instead guide users, help them think critically about government information, and serve as "interpreters of government information" (Schonfeld & Housewright, 2009a).

At one extreme, this led some to imagine government information being provided by "librarians without libraries" where "possession" of information is no longer needed (Shuler, 2008). GPO policy supports just such a model of FDLP libraries by proudly promoting the designation of some FDLs as "digital-only depositories" (GPO, 2015a), later to be called "All or Mostly Online Federal Depository Libraries" (FDLP, 2022a). These are "depositories" into which nothing is deposited. GPO has also described the physical collections of government documents in FDLP libraries as containing "legacy" publications, an adjective appropriated from its use by IT managers to describe software or systems that are outdated and unwanted (Jacobs, 2015). Discussions within GPO and the

FDLP began to center around how FDLP libraries would remain significant without collections. Perhaps users would need librarians more than ever to help them navigate the obscure world of online government information? This culminated in GPO's choice of librarians, not the public, as its Designated Community for its digital repository GOVINFO (PTAB Primary Trustworthy Digital Repository Authorisation Body Ltd, 2018).

The new, digital-age assumption was that there was no need for Federal Depository Libraries to preserve digital information because the federal government took responsibility for preservation. FDLs could point to preserved copies on the government web and had no need to select, store, or preserve copies. But, as a preservation strategy, this would only succeed if the government did indeed commit to preserving all its published Public Information. As described in Part Two, this has not happened.

There was a small but graphic demonstration of the weakness of the "pointing instead of collecting" model in 2013 when NASA took its Technical Report Server offline, making all NASA technical documents, no matter how voluminous or valuable, instantly unavailable (Jacobs, 2013).

The first and second Trump administrations have demonstrated on a much larger scale how political decisions can effectively erase inadequately preserved government information (Jacobs & Jacobs, 2025). Its alterations of informational documents and removal of access to data in 2016 alarmed scientists in particular and prompted small armies of volunteers to rescue data and analyze losses (Dennis, 2016; Gehrke et al., 2021). Groups like EDGI (Environmental Data and Governance Initiative, 2017)—which organized the DataRefuge project (2016) and numerous data rescue events in 2017—worked to collect data and to document the loss of data and the alteration of web-based information. As useful and inspiring as these efforts were, trying to preserve data as it is being altered or removed was not viable a long-term preservation strategy (Lamdan, 2017; Mulligan, 2025). This meant that, when Donald

Trump became president again in 2025, the crisis of loss and erasure repeated itself and prompted EDGI and the Data Rescue Project to once again coordinate data collection efforts (Data Rescue Project, 2025a). These groups coordinated with and worked in parallel to the End of Term Archive project which has archived the .gov and .mil web domains since 2008.

While these important data rescue projects prevented a total loss of information, the fact that they were needed at all points to the pressing need for a Digital Preservation Infrastructure. A DPI is needed to preserve published Public Information as it is released, not as it is being withdrawn. OAIS requires (section 3.1) an archive to obtain "sufficient control" of the information it wishes to preserve. When only one institution (e.g., the federal government) has control of the preserved information, preservation is endangered by a single-point-of-failure problem. To be successful, a DPI must not be vulnerable to bad decisions by any one institution (governmental or non-governmental). To do so, it must provide a way for many institutions to participate in preserving the information that is valuable to their constituencies.

Such a DPI will bring back the integration of collections and services in libraries for communities. It will begin by empowering individual libraries to select, from all published Public Information, information for preservation. Then distributed storage and control of a comprehensive archive of government Public Information will ensure its permanence. The availability of a comprehensive, permanent archive of information will empower libraries to select and define a virtual collection from all preserved Public Information. Libraries will build virtual collections using metadata and create digital services to meet the needs of Designated Communities.

Such a DPI would essentially affirm that the definition of "library" does not change in the digital age. Institutions that select, acquire, organize, and preserve information and provide information services for that information to Designated Communities are, by definition, "libraries."

Institutions that do not fulfill all those roles are not libraries. Institutions that provide services for information selected and controlled by others are analogous to travel agents. Both deal with complex, difficult, disparate, unconnected systems, and help users navigate these systems and find what they need. Just as travel agents have no control over what flights or trips are available or what they cost or what restrictions are placed on them, so such "librarians" would have no control over what information is available or what it costs or what restrictions are placed on its use. This is an unsustainable model for an institution that aspires to be a library (Jacobs, 2009a).

A new DPI must support libraries and archives that provide services for collections that they can select and control.

Addresses and Identifiers

In physical libraries, each volume has a unique identifier ("call number") that serves as its permanent physical address in the library. Although web pages have addresses (URLs), few have permanent addresses (PIDs) or unique identifiers. It is just as essential to have permanent, unique identifiers and addresses in the digital age as it was in the paper era. The paper-era model of a unique identifier serving as an address can work in the digital age, but it will require new methods and technologies.

A Dewey Decimal or Library of Congress call number can uniquely identify a volume by subject, author, and even year of publication. Call numbers are meaningful and unique identifiers because they are based on the identity of the book, not on its physical location. The call number is a unique identifier that also serves as an address. The important distinction is that the identity of the book determines its address, not the other way around. The result is that the physical location of a book can change, but its identity (and its address—its order on the shelf) remains the same.

The web is different. URLs (Uniform Resource Locators) are addresses, but few of them have any useful relationship to the identity of the content they point to. It is much more typical for URLs to conform to the technical constraints of managing a website rather than to the identity of the content. This results in URLs that contain such volatile elements as machine names, the location path of a file, information dependent on a particular Content Management System (CMS), and so forth. All these things change over time, and when they do, links break. This is one of the inherent characteristics of digital information: it is not bound to any single storage medium or location.

The result is "link rot" (links that no longer point to linked information) and "content drift" (content that has been altered at a given URL). These are well-known problems and have been well understood since the earliest days of the web. Tim Berners-Lee wrote about it in 1996 shortly after he created the World Wide Web, and in 1998 he noted that, in theory, there are no reasons at all for URLs to change, but there are "millions of reasons in practice" (Berners-Lee, 1996; Berners-Lee, 1998).

The problem is not limited to obscure sites. GPO's own digital repository has already had four base addresses over the years (gpoaccess.gov, access.gpo.gov, gpo.gov/fdsys, and govinfo.gov). And, as noted earlier, a study by the Chesapeake Digital Preservation Group found link rot for dot-gov websites as high as 51 percent (Chesapeake Digital Preservation Group, 2013), and our own tiny study (Chapter 12) found 85 percent of a sample apparently withdrawn or moved.

One way of dealing with fragile URLs is to create "persistent identifiers" (PIDs). These work like URLs, but they redirect to the current machine address of the data to which they are assigned. This means that a PID never changes and always retrieves the same information even if the digital object is moved or stored in more than one location. There are lots of technologies for creating PIDs, and the choice of PID technology is important, but technology alone does not solve the link rot problem. No technique for constructing URLs is magically permanent. Any link can break. Even GPO's own chosen PID technology (PURL or Persistent Uniform Resource Locator) has itself had three different addresses

(Kraemer & Dahlen, 2018). Regardless of which particular PID technology is used, PIDs require monitoring and maintenance. A solution to link rot requires social commitment and economic resources to implement and manage and maintain the technology over time. PIDs themselves are information, and they require long-term management just like the information they point to does.

PIDs are not just a convenient way of locating information on the web. PIDs are, as a matter of both practicality and functionality, *an essential component of digital preservation*. No PID can be permanent unless the information it points to is preserved, and no information can be considered preserved—and available—if it does not have a permanent ID.

All PIDs are not the same. An infrastructure-scale PID will function both as a unique identifier and as a permanent address. Although this is similar to the way call numbers work in a physical library, PIDs have different and additional functionality. A PID can serve as an identifier in two ways. It can serve as a unique "ID" of content (like a Social Security number serves as a unique ID for a person), and it can provide direct access to descriptive information (such as titles and authors and dates) about the content. A PID works as an address by functioning as an actionable link that directs to the current storage location(s) of one or more copies of a digital object. Current PID technologies such as Handles and ARKs already provide both these functionalities.

The ability to point to more than one location is a significant feature of some PID technologies. It is an essential component of a digital preservation infrastructure that will distribute storage and control to prevent the single-point-of-failure problem described above. It will enable seamless access to digital content during brief service interruptions such as government shutdowns and hardware failures, and prevent loss of information if the government intentionally withdraws or alters information.

In the digital age, our assumptions about addresses and identifiers need to be based on the fact that PIDs are essential, not an optional convenience. "Stored information" without a PID cannot be considered to be "preserved information." PIDs require the same kind of resources for management and maintenance as the information they point to.

A new DPI must support PIDs as essential, not an optional convenience.

Summary

The old assumptions described above are just a few of those left over from the paper-era preservation infrastructure. In Star and Ruhleder's terms, these are learned conventions and practices that are embedded in our existing institutions, social arrangements, and technologies. The new characteristics of digital information demand that we re-evaluate old assumptions if we are to develop new conventions and practices that enable us to successfully preserve government information.

This review of these particular assumptions provides a useful context for imagining some of the elements of a digital preservation infrastructure that we will examine in Part Four.

- Communities. A digital preservation infrastructure can change the way libraries serve their communities: from a one-to-one, geographically based service to a many-to-many, communities-based model that will leverage the power of the library infrastructure to provide more services for more communities.
- Libraries and Archives. Digital-age libraries and archives can cooperate to continue to do those things that define them as libraries and archives while changing *how* they do those things and how they complement each other's services and functions.
- Accessibility. The importance of accessibility as a community-specific function (rather than only a one-size-fits-all "availability") will be paramount to the success of a digital preservation infrastructure.
- Selection and acquisitions. A new infrastructure needs to support both comprehensive archival acquisitions and selective

community-focused collection building. This will enable libraries and archives to address the varied use-needs of Designated Communities both for published Public Information and for unpublished Records.

- Collections and Services. Individual libraries and archives must be able to select and define collections and the functionality of the services they provide for those collections.
- Addresses and identifiers. PIDs are an essential component of digital preservation.

PART FOUR
A Digital Preservation Infrastructure

In Part Four, we offer a vision of a digital preservation infrastructure (DPI) for government information. Star and Ruhleder suggest that infrastructures emerge from the practices of stakeholders. They do not emerge from nothing, but from active choices each library and archive makes today. They emerge by being envisioned, planned, facilitated, nourished, and grown.

We begin by describing the three stages of how institutions adapt to technological change as a context for understanding obstacles that stand in the way of change and strategies that have already been tried (Chapter 18).

Next, we describe elements that a successful DPI must address (Chapter 19). These "elements" are not technologies or methods, but characteristics, functions, values, and responsibilities. This chapter enumerates the roles that need to be fulfilled, a model preservation goal, a planning hierarchy, standards and principles for preservation, and gaps that need to be filled.

With those two chapters as context, we introduce a framework for a Digital Preservation Infrastructure in Chapter 20. This framework comprises a distributed implementation of OAIS functionality and a many-to-many model of providing digital collections and services to users of government information.

Chapter 21 examines a distributed implementation of OAIS from the perspective of the six OAIS functions.

Chapter 22 examines the many-to-many service model from the perspective of the institutions (governmental and non-governmental) that will actually do the work.

Chapter 23 describes the practical transition to a fully functional DPI.

Chapter 24 provides a summary and conclusions of the whole book.

CHAPTER 18
Barriers and Opportunities

A successful Digital Preservation Infrastructure will need successful strategies. Strategies are approaches that institutions use to select their actions and that governments instantiate in laws.

Before choosing new strategies, it can be useful to understand those strategies that have already been tried and where and why they have fallen short. It will also be prudent to acknowledge in advance known obstacles to change. We use a three-stage model of how organizations adapt to technological change to help identify and understand successful and unsuccessful strategies.

Three Stages

A simple, elegant idea that emerged from observations of how organizations adapt to technological change provides a context for evaluating preservation strategies. Clifford Lynch, the executive director of the Coalition for Networked Information (CNI), gave the idea its clearest expression as it relates to libraries (Lynch, 2000). He said that organizations go through three phases in adapting to changes in information technology.

1. Modernization or automation (doing what you are already doing, though more efficiently);
2. Innovation (experimenting with new capabilities that the technology makes possible); and
3. Transformation (fundamentally altering the nature of the organization through those capabilities).

Lynch attributed the idea to Richard West and Peter Lyman, but much the same idea has been expressed by others going back 50 years. Alvin Toffler (Toffler, 1971) used four stages in his book *Future Shock* and noted that the time between each of the stages was getting shorter. John Naisbitt, in his popular book *Megatrends* (Naisbitt, 1982), said that the first stage "follows the line of least resistance" that "is applied in ways that do not threaten people" and that the third stage finds "new directions." Richard Lucier (Lucier, 1996), the University Librarian of UC San Francisco, said that focusing on modernization at the expense of transformation will result in a "modernized traditional library" but not the transformed library needed by users of the 21st century.

Modernization, in Lynch's view, is little more than automating old processes. Libraries have been good at that—converting their card catalogs to OPACs, for example. *Innovation* allows libraries to stretch the capabilities of their OPACs. For example, including more than books and journals in the OPAC: chapters in books, articles in journals, and non-print types of information such as numeric data and GIS data.

But neither modernization nor innovation are sufficient to guarantee the preservation of digital information. *Transformation* is necessary. Lynch concluded that networking our information itself will allow libraries to go beyond the physical limitations imposed by the paper era; it will lead to "a breakdown of geography as an organizing principle of libraries." It is this idea that drives the two approaches that we recommend in Chapter 20.

The three stages of adapting to technology provide a useful terminology for understanding why strategies fail and why obstacles are so persistent. When the inherent characteristics of digital information (as described in Chapter 16) make it difficult to simply modernize an old infrastructure successfully, transformation becomes not a hypothetical possibility for an idealized future, but a practical necessity for today.

What Has Not Worked

As we have noted here and elsewhere (Jacobs, 2014), there are many projects that have successfully saved lots of born-digital government information from permanent loss. But these projects do not aggregate to the comprehensive preservation of published Public Information. As we have shown, there is not even a way to document exactly what has been saved and what has been lost. These many separate projects do not comprise a comprehensive strategy. It is worth examining why that is.

One preservation strategy that has not worked is to assume that the old models from the paper era will work if only they are adjusted to include digital information formats. As we noted in Chapter 5, changes to laws and policies have attempted to do this without success.

Another prominent example is that of GPO's approach of offering support to agencies for the "entire publishing lifecycle of tangible and digital information" (GPO, 2023d). This is an attempt to "modernize" publishing. It is tempting to wish that all federal government agencies would simply use the Government "Publishing" Office the way they once used the Government "Printing" Office in the paper era, resulting in preservation as a byproduct of publishing. This approach can work in the digital age when two factors merge: when information is issued as traditional publications and when the producer is willing to cooperate with GPO. This is most evident in the case of Congress because much of its Public Information (e.g., bills, acts, hearings, reports) is still issued as traditional publications and because GPO is part of the legislative branch and thus has close ties to Congress. But, when either or both of those factors are missing, the model breaks. It has not made a significant dent in digital preservation of the executive branch because much of its Public Information is published as web-content, not as traditional "publications," and because agencies are not cooperating with GPO. They are not cooperating with GPO because the old infrastructure essentially eroded away. The 2018 Library of Congress study documented this erosion. It found that the staffs of federal agencies had limited knowledge of GPO's digital product dissemination and preservation missions. This was caused by the shift to digital publishing, which resulted in a loss of institutional knowledge as print-era staff (printing officers, librarians, archivists) were phased out (US Library of Congress, Federal Research Division, 2018).

Another strategy is to create a new digital government publishing infrastructure that would include preservation as a byproduct of publication. Such a strategy would surely classify as an attempt to "innovate" publishing, but attempts at innovating publishing have not helped preservation significantly. Advances in coordinated digital publishing by GPO (the Federal Publishing Council) and other agencies (US Web Design Systems, US Digital Services) are not yet focused on preservation at all, much less on recreating a centralized preservation infrastructure of the past (GPO, 2023b; digital.gov, 2023; US Digital Services, 2023).

It is also evident that individual isolated preservation projects, no matter how useful they are, have not succeeded in preserving the enormous amount of digital government information being produced. GPO and NARA and the Library of Congress have all acknowledged that collaboration is needed, but those efforts have, so far, been inadequate. As shown in Part Two, they have resulted in unnecessary and inefficient duplication, large gaps in preservation, and not enough innovation in discovery, accessibility, and utility. While the shared responsibility strategy worked in the paper era, digital information has broken the model. The sheer quantity of information produced in the digital age has overwhelmed the preservation agencies. GPO and NARA and LC have each justified their own limited preservation of executive branch information by hoping that the other two will preserve what they cannot, but that has not happened. The inherent characteristics of digital information have blurred the distinctions between Records and Publications, leaving the "modernized" laws inadequate to deal with either type of government information (Chapter 5).

Government-led collaboration projects have, so far, had limited impact on preservation. The Federal Web Archiving Group has not solved the problems of too-much-to-harvest and not-enough-resources (Neubert, 2015; US Library of Congress, 2017; GPO, 2017). The standards promulgated by the Federal Agencies Digital Guidelines Initiative (FADGI), as good as they are, have not had wide enough effect (Federal Agencies Digital Guidelines Initiative, 2023; Bethany, 2018). And GPO's vision of a "National Collection" acknowledges the necessity of regional- and national-level collaborations but focuses FDLP efforts on geographically constrained "local level" access, last-generation online catalogs tied to local libraries, and the antiquated *Catalog of Government Publications* (FDLP, 2021b). Its vision does include some nascent ideas (preservation stewards, digital content contributors, and digital access partners), but so far, the focus of these appears to be on preserving printed documents or digitization of printed documents (FDLP, 2023b). Here, resources are going to information that is already preserved and accessible (50 projects) rather than to born-digital information that is being lost every day (one project).

Lessons Learned

These projects reveal three important lessons for the preservation of born-digital government information. First, these inadequate strategies are all minor modernizations of existing print-era strategies. Storage and availability have been modernized, but there is little innovation in access and no transformation in services. Second, comprehensive preservation cannot rely on isolated projects. Third, the existing laws, regulations, and policies of the federal government are inadequate for comprehensive preservation of the government's Public Information. Indeed, focusing on modernizing old laws may actually delay focusing on transformative actions that are needed to preserve digital information.

Barriers to Change

Developing a digital preservation infrastructure will require change, and change is almost always difficult. Indeed, Star and Ruhleder point out and document that existing infrastructures themselves can be barriers to change. Existing standards and practices can be rigid, and organizations and staff may resist change. One of the great strengths of libraries and archives is that they are stable and last a long time, but that very stability can slow down or even obstruct evolution. When photographic microfiche was introduced as a medium of distribution of government information in the late 1960s and early 1970s, for example, it broke what Star and Ruhleder would call the transparency of the print-based infrastructure, and it took seven years for the depository program to adapt (Schwarzkopf, 1978). That change was minor when compared to the digital shift. Keeping up and adapting to the rapid changes in the creation, distribution, and use of digital information presents a much more difficult problem for libraries and archives. As we outlined in Chapter 9, it took GPO 18 years to adapt to web-based publishing by agencies, and to do so GPO had to reverse a long-standing policy. In the face of the rapid evolution of digital publishing using the web, this slow rate of change is a barrier to preservation.

The gaps identified in Part Two suggest to us that, although the government has essential roles to play in preserving its information, it cannot do so alone. Preserving all the government information that deserves preservation will require many participants from both inside and outside of government. This will require a strategy with a national scale that bridges institutional boundaries. Planning for the preservation of digital government information will not happen in a vacuum, either. Approaches that are specific to preserving government information will take place in the context of and will overlap with the preservation of digital information in general. In 2018, Oya Rieger interviewed digital preservation "thought leaders" (directors, deans, program managers, executive directors, and other high-level managers) for an Ithaka S+R study of the state of digital preservation (Rieger, 2018). Rieger identified 16 "challenges and gaps" and suggested three overarching issues worth exploring. Several of the barriers to preservation Rieger describes overlap implicitly with the gaps we found for the preservation of government information. Some are simply barriers to change itself.

The focus of Rieger's study was on resources needed for scholarship in general, including cultural, historic, and scientific information, but many of her findings apply to government information. For example, the thought leaders identified open-access content and primary sources as being vulnerable to loss. Government information falls squarely in both these categories. Several interviewees explicitly noted the need to preserve government websites taken down due to politicized rhetoric. (We would suggest that this is an essential example of a broader principle: that comprehensive preservation of Public Information is necessary to document changes in and evolution of government policies [Jacobs & Jacobs, 2025].) Similarly, the report notes that worrying about new emergent content formats can mask the fact that existing projects have failed to completely preserve valuable older, simpler, text-based content in ejournals, ebooks, and electronic theses and dissertations. We would add to that list the text-based information that comprises the vast majority of government information (see Chapter 6) and question (as NARA does [NARA, 2008c]) the degree to which preserving the look and feel of websites is an adequate way to preserve Public Information.

Rieger notes that lack of assessment metrics is a key problem, paralleling the preservation-information gap and the inventory gap that we described in Chapter 13. She quotes Clifford Lynch as describing digital preservation as being "a metric that has defied measurement."

Perhaps the biggest barrier that Rieger identifies is that preservation is no longer an unquestioned moral imperative for library leaders. This is exactly why we are emphasizing the Star and Ruhleder view of infrastructures and the essential need for a culture of preservation as a prerequisite to a successful DPI. Rieger goes further, identifying institutional politics as more of a hinderance to preservation than technical issues. She says that this situation means that preservationists have to demonstrate to management the value of preservation. A more recent study, a systematic literature review of trends toward in-house digital preservation, suggests that this attitude may be changing (Ahmad et al., 2023). But preservationists will recognize that they still must, as Rieger says, "convince provosts" and "make the case" for the importance of preservation to libraries' missions. As Courtney Mumma, Deputy Director of the Texas Digital Library, said as recently as 2023, preservation is still undervalued and preservationists still have to "convince executives, owners, donors, granting bodies, presidents, provosts, legislatures, technologists, managers, and *even other librarians and archivists*, that digital preservation work is important and essential" [emphasis in original] (Mumma, 2023).

We might identify this issue as another gap: a gap in responsibility for preservation. That links to the concern that there is too much to preserve (see the Introduction to Part Two) and the acceptance of a responsibility for access—but not for preservation (Chapter 17). We may hope that more library managers are recognizing that, as Ahmad says, "preservation is an intrinsic task," but regardless of how many already do, it is undeniable that this is an essential element of a successful DPI, and its absence is a barrier to successful preservation.

Another key issue that Rieger identified is the ownership and control of information. As one respondent said, "It is difficult to preserve content that is not 'owned' or 'controlled' by libraries." While commercial ownership of content is often cited as an obstacle to preservation, it is one obstacle that government information mostly avoids. Since most federal government Public Information is not owned or copyrighted by its creators, it is possible for preservationists to acquire and control it without the hinderances that exist for commercial, licensed, and copyrighted content. (Indeed, libraries do not need a law granting them permission to preserve government information. This basic fact underlies the existence of the Internet Archive and the EOT harvests.) But *control* is still an issue for government information, particularly when the responsibility for preservation is relegated to the government alone. When the only preserved copy is controlled by the agency that produced it and that agency has no legal mandate or budget for preservation, preservation is put at risk. The agency that produces information has its own perspective on what is worth preserving and what can be discarded, and that perspective may not reflect the needs of all users. For information that is preserved solely by the government, its preservation is at the mercy of political and funding decisions of that government.

"Web archiving" as a preservation technique does not solve the preservation problem, either. Rieger points out that users have questions about web archives because they do not have a clear understanding about "how archived websites are discovered, used, and referenced." From our perspective, this is a basic problem of the current state of web archiving: it saves websites. But we believe that most users do not want to browse or use old websites; they want the information those websites held. But the current technologies for presenting the contents of web archives forces the user to get to the information they want by navigating through the preserved version of the old websites. Sometimes the functionality of those old websites cannot be preserved, making the information difficult or impossible to retrieve. When the only preserved copy of a document is in a web archive that can only offer limited ways that the information can be identified, discovered, accessed, and used, preservation is hindered.

Lessons Learned

The challenges that Rieger identified may be discouraging, but we believe they provide clues to how a Digital Preservation Infrastructure can succeed. For example:

- Library managers value "catering to the university's immediate needs." Those managers who no longer see preservation as an unquestioned moral imperative may be persuaded of the need for preservation by connecting it to those immediate needs. This can be done by explicitly defining Preservation as including accessibility and usability for current users as well as future users (Chapter 4). Managers must be led to understand that, in the digital age, the future comes as quickly as a file can be moved or altered or erased. "The future" arrives this afternoon.
- Because one of the challenges of web archives is a neglect of an emphasis on how archived content will be discovered, accessed, and used, web harvesting should focus more attention on those issues. A successful DPI should focus as much on discovery and delivery as on harvesting and storing.

Rieger also has specific suggestions that could guide a vision of a successful DPI:

- Preservationists need to have a common understanding of the roles of libraries and archives.
- There is a need for a policy infrastructure and "overall strategies" and "a cohesive and compelling roadmap" for preservation.
- The roadmap would require the development of a shared understanding of what digital preservation comprises.

As we will demonstrate in the remainder of this part, creating a shared goal and developing a goal-based long-term plan for preservation would address all three of these suggestions.

Preservation Paradoxes

Drawing on our definition of Preservation as including accessibility and usability of information, we must add a few points to those above. A successful DPI will have to accommodate seemingly paradoxical requirements. For example, the DPI needs to:

- accommodate the comprehensiveness of archives and the selectiveness of libraries;
- empower archives to select Records for their archival value (historically 1 percent to 3 percent of all Records) while empowering libraries to select versioned published Public Information (closer to 100 percent) as documentation of changing public policies;
- preserve provenance and original context while enhancing discoverability, access, and use, based on subject, discipline, geography, use-case, etc.;
- preserve information once, but make the same information accessible in different ways to accommodate the different needs of different Designated Communities;
- accommodate small, focused collections without creating isolated silos of information;

- develop long-term strategies that empower short-term, fast-changing tactics;
- employ strategies that work within the context of inadequate preservation laws without being inhibited by those laws;
- preserve mementos of information that changes unpredictably.

Conclusions

This leads us to some preliminary conclusions that can inform an agenda for a DPI.

- The primary focus of preservation is not just ensuring that a copy is stored somewhere, but that the *use-needs of Designated Communities* are met.
- A successful DPI will require two essential elements that have, so far, been lacking: a shared, well-articulated goal for digital preservation and a strategic plan for reaching the goal. These two elements alone would help address Rieger's suggestions about a "shared understanding" of preservation and a "roadmap" for preservation. But they go further. A well-defined, comprehensive preservation goal can provide a common focus for all projects big and small. A comprehensive, goal-focused preservation plan will enable prioritization and interoperability and flexibility of all projects big and small. Together, they can transform preservation.
- A preservation plan that enables libraries to select and preserve content and control it for the delivery of services that match the needs of their communities will be more likely to have the support of library managers—and their communities—both now and in the future. This will help define a "common understanding" of the role of libraries. Libraries will be more than "travel agents" to the web; they will actively collect, preserve and provide services for information their communities need.

- To address how archived content will be discovered, accessed, and used, preservationists need to examine their use of web harvesting. Because user interfaces to archived web content are still rather primitive, web-archiving technologies should be updated to enhance discovery, delivery, readability, understandability, and usability by Designated Communities. In the particular context of government information, this also includes a robust interface to different versions (mementos) of the same content (Van de Sompel et al., 2009). Web harvesting itself (which attempts to preserve websites) should, perhaps, be reimagined as "information harvesting" (which would preserve and make accessible the information that websites make available).

- For preservation planning to be "cohesive," preservation projects should be built on a common infrastructure that accommodates the integration of services with collections across institutions. Specialized collections and services that are integrated with other services and a comprehensive collection of government information should be preferred over isolated silos of content.

- Preservation planning should explicitly accommodate the "immediate needs" (i.e., the information service needs) of Designated Communities today and be flexible enough to adapt to the needs of users in the future (Johnson & Kubas, 2018).

- The focus of preservation should be on content and people using the content.

CHAPTER 19
Elements of an Infrastructure

In this chapter, our goal is to specify essential elements of a Digital Preservation Infrastructure. A DPI will comprise shared norms that define a culture of preservation. In this context, "culture" can be understood literally as a society of shared values and traditions and habits, and metaphorically as the biological ground of a healthy ecological system in which preservation flourishes. A culture cannot be designed and built like the physical substrate of an infrastructure can be. A culture emerges from how a physical infrastructure is used. But the culture of a successful DPI will have effects and outcomes that can be planned and predicted.

A vision of such an infrastructure that supports such a culture must begin with an understanding of the elements of the ecosystem. Some of these elements are physical substrate; some are ideas and values and methods and roles. Some elements can only be expressed as problems that must be addressed. Even though these elements are very different kinds of things, they can all be specified in broad terms in order to help define the landscape of preservation within which the culture and the infrastructure will emerge.

The elements we describe here are, for the most part, not new and should not surprise preservationists. Most of what we assemble here is based on OAIS, the 2002 NDIIPP report *Preserving our digital heritage: plan for the National Digital Information Infrastructure and Preservation Program*, and our own research as presented in parts 1-3 of this book. (ISO, 2012b; National Digital Information Infrastructure and Preservation Program [US], 2002).

The elements of a successful DPI ecosystem will allow it to:
- Use goal focused planning
- Follow preservation standards
- Follow principles

- Fulfill the traditional roles of libraries and archives
- Fill known gaps in preservation
- Resolve the apparent paradoxes of preserving government information for use
- Address user needs

Goal-Focused Planning

A successful DPI must have a common goal: the preservation of government information. As OAIS says, "the primary goal" of an OAIS compatible archive is "to preserve information for a designated community over an indefinite period of time." This goal will guide all preservation decisions and will enable planners to develop strategies, prioritize objectives, and evaluate and choose tasks and tools.

A successful goal and planning process will ensure that the projects chosen will be efficient and effective. Although the specific goal and planning process will have to be defined by preservationists, we offer, as an example, our own view of a well-defined goal and an outline of a planning process here.

Defining the Goal

A goal of preserving government information is literally meaningless if it does not specify what "preserving" "government information" means. The National Digital Information Infrastructure and Preservation Program (NDIIPP) described its goal for digital preservation as encouraging shared responsibility and seeking national solutions that would embrace collection, selection, and organization, long-term storage, preservation, authenticity, and persistent access. A successful DPI for digital government information will require a common understanding of what a community of preservationists wish to preserve, why they wish to preserve it, and what it means to preserve it. We presented our own definitions of these elements in Part One.

"Information" or *"Information-as-thing"* means the object, such as a book or a digital object, that is used for the transmission, storage and use of "information-as-knowledge." The Information that a DPI will preserve is the digital objects that can be transmitted, stored and used.

The target of preservation is "Public Information"

'Public information' means any information, regardless of form or format, that an agency discloses, disseminates, or makes available to the public [44 USC 3502; OMB 68]

Public Information includes what we traditionally called "publications"—information packaged for the public and for public dissemination—as well as those Records that have been released by an agency without restrictions on access or use by the public. Such Records would include Records released under FOIA mandates, and most of those that are transferred by agencies to NARA. It might also include research articles and data funded by government grants. Many of those Records will not have been neatly packaged for public use like publications are, but as soon as they are available for use by the public, they become, by definition, "Public Information" and, therefore, targets for preservation. In other words, a successful DPI will accommodate what we might have traditionally called "library materials" as well as "archival materials."

It is necessary to preserve this information for three reasons:

- Government information has inherent value as the official record of the actions of a government and so must be preserved as an essential prerequisite to informed public participation in a democracy.
- The information that the government collects, compiles, aggregates, and creates in the course of its normal functions must be preserved because it comprises irreplaceable knowledge about and for the republic.

- The inherent value of government information is realized through its accessibility and usability by the general public.

Preservation must be understood as more than saving stuff and storing it. For information to be "preserved," it must meet the OAIS functional criteria, which we summarize in our definition of preservation:

> Preservation means maintaining information for discovery, delivery, readability, understandability and usability by a Designated Community for the long-term.

The Planning Hierarchy

Successful preservation of government information will require serious planning that enables and empowers many collaborators to easily work toward a common goal without the encumbrance of unnecessary bureaucracy. Preservation planning is one of the six functional requirements of OAIS, which defines planning as developing preservation strategies and standards.

One useful way to manage planning is to divide the process into a hierarchy of types of activities. Such a hierarchy can help keep a plan focused on an overarching goal while allowing the flexibility that is needed to respond to changing conditions and opportunities. We suggest a specific hierarchy of planning levels as a useful template for long-range, big-project planning:

—— Goal
——— Strategies
———— Objectives
————— Tasks (tactics)
—————— Tools (technologies)

Each of these five levels has a hierarchical decision-making relationship to the other levels. Every decision at any level of the hierarchy is made to fulfill the intentions of the level above it. Thus, Strategies are chosen to achieve the Goal, Objectives are established based on those Strategies, and so forth. Any decisions that cannot be justified up and down the hierarchy are outside the plan.

Decisions change more quickly at lower levels of the hierarchy and more slowly at higher levels. Strategies change slowly and with more deliberation because they affect everything below them in the hierarchy. Tools at the bottom of the hierarchy can be replaced quickly to meet changing conditions. But even tools must be accountable up the hierarchy all the way to the Goal.

Using such a goal-focused planning process provides both long-term and short-term flexibility. For the long term, it will accommodate new types of information, new user communities, new uses of content, and the availability of new tools. For the short term, it can adapt quickly to different types of digital Public Information that can be best preserved with different tactics or tools. The hierarchy can branch at any level, using different methods to achieve the same intent. Goal-focused planning accommodates several strategies, many objectives, lots of tasks, and innumerable tools.

Finally, explicit, transparent goals and strategies provide a framework for projects of all kinds and sizes to develop organically with a minimum of bureaucracy. In this environment, bureaucracy is a tool at the bottom of the hierarchy, not a strategy or an objective. This accommodates the participation of small partners with limited resources as well as large ones with more resources. It encourages the building of systems that link together naturally as part of a large-scale preservation infrastructure and helps avoid the building of isolated silos of information.

Preservation Standards

Star and Ruhleder say that one of the characteristics of infrastructures is the embodiment of standards. Our brief, simple definition of Preservation is derived from OAIS and needs to be understood in the context of the OAIS preservation standard. A successful DPI should, at a minimum, "embody" these standards by ensuring that all the OAIS functions are performed and all OAIS responsibilities are met for all preserved information.

OAIS Functions

1. Preservation Planning
2. Ingest
3. Data Management
4. Archival Storage
5. Administration
6. Access

OAIS Responsibilities

1. Negotiates for and Accepts Information
2. Obtains Sufficient Control for Preservation
3. Determines Designated Community
4. Ensures Information is Independently Understandable
5. Follows Established Preservation Policies and Procedures
6. Makes the Information Available

Guarantees

Our own summary of the OAIS standard is that, in order to claim that information is being successfully preserved, a preservation must provide five guarantees. Information must be:

- Not just preserved, but discoverable. [2.2.2]
- Not just discoverable, but deliverable. [2.3.3]
- Not just deliverable as bits, but readable. [2.2.1]
- Not just readable, but understandable. [2.2.1]
- Not just understandable, but usable. [4.1.1.5]

Principles

One of the lessons we take from the many attempts to codify principles of preserving government information (Chapter 1) is that there *are* underlying principles that a DPI should honor. Therefore, a DPI must guarantee:

- privacy for users
- transparency of methods and values
- authenticity of the information preserved
- free accessibility of the information preserved

These are not goals or methods; these are guardrails within which planning operates. The goal-focused planning must ensure that projects and tools that violate these principles be modified so that they comply with these principles or be rejected if they cannot.

Roles

The traditional roles of libraries and archives are to select, acquire, organize, and preserve information and to provide information services for that preserved information including delivering it to their chosen communities (Borgman, 2003). We suggest that these roles remain crucial to users in the digital age. Preservation will not be successful unless these roles are fulfilled by trusted, reliable institutions that have a primary mission of ensuring long-term, free, public access to government information.

The OAIS functions and responsibilities parallel those traditional library and archival roles. This is a technical way of saying that a DPI need not change *what* libraries and archives do, but it opens up *how* they do so.

(We do not agree with those who argue that the roles of libraries have changed, or should change, in the digital age. As we explained in Chapter 17, those who have argued that libraries can function as service centers without collections have misinterpreted the lessons of those who advocate for more "shared collections." Using paper-era discovery tools to point at "publications" whose defining characteristics, identity, existence, address, and accessibility are controlled by others is simply inadequate in the digital age.)

We believe that it is self-evident that, in order to provide information services, a library must have sufficient control of the information for which it provides services. Obtaining sufficient control of information is one of the six OAIS mandatory responsibilities. This is even more important in the digital age when those who control information also control the technologies that define how it can be discovered, accessed, delivered, and used. Pointing to a digital object is only the last phase (availability) of making information in that digital object accessible. While a basic level of availability of originally archived digital objects will be sufficient control for some services, more advanced and specialized services will require additional control in order to provide additional methods of discovery and delivery of information.

Fill Gaps

One measure of the success of a digital preservation plan will be that it fills existing gaps without creating new ones. In Part Two we identified six gaps.

1. inventory
2. preservation
3. legal
4. infrastructure
5. preservation information
6. preservation planning

Resolve Paradoxes

A successful DPI must be able to resolve the apparent paradoxes we listed in Chapter 18. This will require the DPI to be flexible in a number of ways.

Given the nature of government information, the inherent characteristics of digital information and the need to preserve Publications and Records, the infrastructure must be flexible enough to cope with a changing technological environment over which it has no control. It will need to accommodate different and changing procedures and methods and technologies. It will need to address the specific preservation needs of different kinds of information and the needs of different kinds of communities. It will need to accommodate small and large institutions. The planning process must accommodate flexibility on both the micro-level (e.g., the tactics and tools used by individual projects) and on the macro-level (i.e., the strategies and objectives used to meet the overall goal). The DPI should avoid an inflexible, one-size-fits-all model of preservation. Much like OAIS, it should embrace standards for functionality without limiting how that functionality is achieved.

Address User Needs

The most important element that a successful DPI must address is the information needs of users. Since the purpose of preservation is to provide usable information for users now and in the future, a DPI simply cannot be judged as successful unless it functions for users. A system that is designed to function only for librarians and archivists will impose an unwanted layer of intermediation between users and information. A system that enables librarians and archivists to build digital services for Designated Communities is a system that completes the use-requirements of Preservation.

Although we have repeatedly referred to general user information needs, there are two specific elements that need to be addressed explicitly: the selection of content for preservation and making that preserved information available.

Selection

Users want the information that they need to have been preserved and made accessible. How can a DPI ensure that all the information that users will want in the future is preserved before it is deleted or altered? This question does not have an obvious answer for digital government information for several reasons. One reason is that selecting what to preserve is quite complex, particularly when canonical text is issued in more than one presentation and more than one digital format. Executive Orders, for example, are often issued as at least four different web page formats and three differently formatted PDF files (see Appendix C). Another difficulty is the sheer volume of Public Information. It is essential to preserve government information "comprehensively" in order to preserve its context and provenance and in order to preserve a record of changing government policies. But what methods can be used to adequately identify, appraise, and choose among millions of PDF files and hundreds of millions of web pages, not to mention millions of other files in hundreds of other formats? Finally, if a DPI successfully builds an enormous comprehensive collection of Public Information, how will users find the needle they need in a warehouse full of haystacks?

A successful DPI selection strategy has to be precise in order to get the content users need in the formats they require. It also has to be broad in order to preserve comprehensively. It can do both by engaging many selectors, each reflecting one or more specific Designated Communities.

Availability

To say that users want preserved information to be discoverable and acquirable and usable should be a given for any successful DPI, but saying so can inadvertently mask a simple underlying user-need. Specifically, users want information to be freely available permanently and without interruption. As Judith Russell, the Superintendent of Documents from 2003 to 2007, has said, in a digital age, an interruption of availability "is equivalent to locking the doors of the Libraries"

(Russell, 2010). Assuring availability of preserved information will, we believe, require two elements of any DPI: sufficient redundancy and Persistent Identifiers.

A successful DPI must have enough redundancy to avoid single points of failure. A successful DPI must guarantee the free availability of information even under conditions that could cause an interruption or cessation of services. Such conditions include technological problems, funding shortfalls, changing government policies and priorities, and government shutdowns.

As noted in Chapter 17, the inherent characteristics of digital information make Persistent Identifiers (PIDs) an essential component of digital preservation. No information can be considered preserved if it does not have a Persistent Identifier.

Together, guaranteed availability and PIDs will provide a foundation on which more complex accessibility and services can be built with confidence.

One might summarize these user needs by saying that a successful DPI will:

- preserve government information comprehensively;
- ensure the uninterrupted free availability of preserved information;
- provide defined collections to meet the needs of specific communities;
- provide information services designed for specific communities.

CHAPTER 20

An Open Framework for Digital Preservation

In this chapter, we introduce a framework for a Digital Preservation Infrastructure (DPI) for federal government Public Information.

We are not describing an engineering solution because no one can accurately predict the specific technologies needed to cope with the future of information, digital formats, user communities, evolving information uses, and so forth. Neither are we describing a complete infrastructure in the Star and Ruhleder sense. Rather, we are describing a two-part framework that can serve as a foundation for the growth of an infrastructure. We call this a "framework" because it is a conceptual approach to preservation within which consistent strategies can be developed. It is an "open framework" because it is simply an idea that can be adopted freely.

The framework provides a foundation for growth of an infrastructure by defining the relationships that preservationists can use to work cooperatively toward a common goal of preserving federal government Public Information. For this task we draw on the history, data, and ideas presented in Parts One to Three of this book and the elements listed in Chapter 19.

This framework provides a foundation for a functional specification that can be implemented in many ways. In this way, it is analogous to the way that OAIS provides standards for functionality without limiting how that functionality is implemented. Implementation needs to accommodate many institutions of different sizes and resources that support different kinds of communities. It needs to be able to adapt to changing information content and methods of distribution, changing formats, changing uses, and evolving communities of use. This framework can also be considered a template for developing the strategies of a goal-driven planning process. Engineering solutions are about tools used to implement strategies and accomplish goals. Engineering solutions and specific technologies are closer to the bottom of the planning hierarchy; this framework provides a context for developing strategies at the top of the planning hierarchy (Chapter 19).

Two Approaches

Using the terminology of the three stages of technological adaptation from Chapter 18, we take as a given that a successful DPI will need to involve a transformation of how libraries and archives select, acquire, organize, and preserve information and how they provide information services. The framework needs to do more than automate old processes and more than introduce innovative services. It must transform how libraries and archives preserve information and provide services to their communities. It does not change what libraries and archives do but how they do it.

We suggest that a successful DPI can evolve from a framework that uses two transformative approaches to preservation and services.

- A distributed implementation of OAIS
- A many-to-many model of collections and services

Libraries and archives will transform when we stop thinking of each one as an isolated institution and start thinking of all of them as parts of a single information ecosystem. Instead of thinking of each library as serving only a small, geographically bound community, we can think of libraries as serving Designated Communities wherever the members of those communities live and work. Instead of thinking of small, geographically localized communities being served by a single local library, we can think of each Designated Community being served by libraries everywhere. Each local library can develop services for its own constituency and provide those digital services to users everywhere though a global infrastructure. Where we used to think of OAIS as applying to one institution at a time, now we can think of many institutions collectively comprising an Open Archival Information System. Where we once thought of libraries and archives as separate but complementary, we can start thinking of them instead as integral parts of a whole. Where once we thought of libraries and archives collaborating through the sharing of their isolated collections, we can think of them as collectively preserving a common collection and making it available and accessible universally.

OAIS

We envision a distributed implementation of OAIS for two practical reasons.

First, a thoughtful distribution of responsibilities will increase day-to-day reliability of the system and eliminate the single-point-of–failure problem. A system with a single point of failure is vulnerable to interruptions of services and loss of collections when that single point fails. Such failures can be short term or long term, and intentional or unintentional. They can be technical, political, administrative, or budgetary. A distributed system has enough redundancy to prevent any single failure from affecting either the integrity of the collection or the delivery of services. Notably, it also prevents any intentional political decision to alter or remove information from public access once it has been archived.

Second, the scale of the task of preserving government information will require the participation and cooperation of many institutions, many selectors, many service providers, and the definition of many sub-collections. The high-level planning of a distributed implementation of OAIS can provide guidance for many disparate projects to target gaps, reduce unnecessary duplication, and avoid isolating information in "silos." Planning based on standards and shared strategies will make participation easier and reduce the need for legal contracts and bureaucratic overhead and simplify their use when they are desired.

The need for cooperation is well recognized philosophically (Hirtle, 2008; FDLP, 2021b; GPO, 2017; NARA, 2008a; US Library of Congress, 2017), but it needs a better practical implementation than has so far been developed by the digital preservation community.

Using a distributed implementation of OAIS is not a new idea (Fenton, 2011; Lavoie, 2014; Neto et al., 2017; Spence, 2006; Zierau, 2017). Martha Anderson of the National Digital Information Infrastructure and Preservation Program (NDIIPP) described OAIS as "an ecosystem of different actors and processes across the life of digital content. Each actor contributes to the whole in unique fashion; each part of the process influences another" (Anderson, 2011a). There are ample examples of projects that distribute OAIS functions in different ways. These include the California Digital Library (Abrams et al., 2010), the Canadian Association of Research Libraries Portage Network (Qasim et al., 2018), the Cedars project (Jones, 2002), the MetaArchive (Schultz, 2010), and the PLANETS project (Planets Preservation and Long-term Access through NETworked Services, 2007). Indeed, GPO already distributes parts of the OAIS Archival Storage function (error checking, media refreshing, disaster recovery) of its GOVINFO repository to 36 partners of the LOCKSS USDOCS program (LOCKSS Program, 2022).

In Chapter 21 we outline a practical, efficient way of distributing OAIS responsibilities.

From One to Many

Traditionally, libraries use a one-to-one model of collections and services, but we envision a transformation—for government information—to a many-to-many model. With the one-to-one model, each individual library serves one local constituency. The many-to-many model uses the inherent characteristics of digital information to break the limitations of providing physical books to nearby users. It replaces geography as the organizing principle of libraries and enables many libraries to serve many communities everywhere. It is user-focused and makes it easy for users anywhere to find and use collections and services developed everywhere.

One-to-One

Providing services to a local constituency is the greatest strength of the one-to-one model but also its greatest weakness.

It is a strength because each library has a well-defined community and the library and the community have direct, consistent contact. The library supports the information needs of the community, and the community supports the creation and maintenance of the library. The library is part of the community and reflects the values and needs of the community. It also accommodates the physical constraints of the print era, literally putting each volume within reach. The size and scope of both the community and the library are well defined. All this makes it relatively straightforward for a library to develop collections and services that meet the needs of its community.

The biggest weakness of the model is that it tends to force a library to treat all of its constituents as a single user community. The community may be large and complex, but it is the library's role to address the needs of the community as a whole. This works to an extent because each library's community does have its own common identity. For example, each school and college and university has its own educational goals. A public library serves a literal community such as a state or county or city or neighborhood. But the commonalities of such communities are general and do not account for the more specialized information interests and needs and uses of sub-communities—communities of interest. Of course, libraries understand this, but resources are rarely adequate to serve more than a small handful of sub-communities. The physical constraints of the print era also limited the methods of addressing those needs. Typically, libraries use two common methods to alleviate the limitations of the one-to-one model.

The first method is inter-library loan (ILL). In a sense, ILL is one way of implementing a many-to-many library model. ILL makes the collections of each library available to all libraries and the collections of all libraries available to each library. Unfortunately, constraints limit the effectiveness of ILL as a many-to-many model. One constraint is that it is slow and costly, which limits its use and utility and scope (Gaffney & Massie, 2022; Link et al., 2015). A second constraint on ILL is that it tends to share only collections and not services. Although users of one library can, technically, borrow books and articles from almost any library, to do so they must identify the information that meets their needs and that requires information services. As one study of two university libraries noted, ILL requests "require some effort on the part of the faculty or student and incur some delay," and that creates "a brake on incidental or accidental use of content" (Nabe & Fowler, 2015). Both constraints have been reduced with the increasing availability of online information services. These include the sharing of online bibliographic information, the use of the web services such as LibGuides for collection navigation (Springshare, 2023), the digital automation of ILL requests, and the ability to share digital copies. But ILL is an old model built for the paper era. In Lynch's terms, it has been modernized through automation, but there has been little innovation and no real transformation. Neither the library nor its ILL functions have been fundamentally altered through new capabilities.

The second method libraries use to alleviate the limitations of the one-to-one model is to create specialized collections and services specifically for sub-communities. There are many examples of this, the most prominent of which are law libraries and medical school libraries on university campuses. Some large university libraries have many specialized libraries; the University of Illinois has 41, Harvard has 28, UC Berkeley has 22, UT Austin has 19 (University of Illinois, 2023; Harvard Library, 2023; University of California Berkeley Library, 2023; University of Texas at Austin, 2023). Similarly, large public library systems can tailor the collections and services of branch libraries to their neighborhoods and often have specialized collections and services (e.g., for children) within branches. Many libraries also have specialized services for defined collections such as a serials, music, microforms, data and GIS, and, of course, "government documents." All of these examples have two elements in common: a large enough community and a large enough defined collection. The sub-community that needs a special service must be large enough to warrant a library investing its limited resources into providing the collection and service, and the defined collection must be large enough to meet the needs of the sub-community.

This makes it particularly hard to implement for smaller libraries, for smaller collections, and for smaller communities. Defined collections may be based on subject matter (law, medicine), demographics (an undergraduate library on a university campus, a children's library inside a public library), format (microforms, recorded music), and on how the collection is used (data analysis, creation of maps). Traditionally these services are provided by staff who specialize in the subject, community, format, or use. The existence of such services in libraries today demonstrates their utility. Providing a specialize service for a defined collection to a designated community of users can be better for users than lumping all collections into a single, one-size-fits-all service.

Many-to-Many

The many-to-many library model contrasts markedly from the one-to-one model. Implementing it will require the kind of transformation that Clifford Lynch describes, and it will require the kind of new social norms that Star and Ruhleder describe. It empowers all libraries to make their collections of government information available to all users everywhere though their digital information services.

The many-to-many model does not replace the one-to-one model but adds to it in three ways. First, some collections and services will continue to be provided best at the local, in-person, constituency level.

Second, the many-to-many services a library provides will serve its own community *plus* (not instead of) similar communities everywhere. Third, each library will be able to provide, to their own local communities, digital services and collections developed by all libraries.

In this context, a library's local constituency will be served, not only by its own local library, but by many libraries that are providing compatible services to their own chosen Designated Communities. Libraries that once could afford only one or two defined collections with specialized services can still provide those, but their community will have full access to all the collections (comprehensive and sub-collections) and services provided by all the other libraries and archives participating in the many-to-many model. By using the inherent characteristics of digital information, the many-to-many model can, in Lynch's terms, replace geography as the only organizing principle of libraries.

The many-to-many model transforms libraries from serving only a local constituency to serving Designated Communities. Libraries will be able to choose which Designated Communities to serve much the way libraries chose which sub-communities to serve in the paper era. Libraries with similar community-needs will be able to collaborate in the development of collections and services for that community. The collection for a Designated Community might be based on subject or discipline (economics or history), or format of information (file type or media type), or provenance (agency or office), or on how the information is used (close reading, distant reading, computational analysis), and so forth. Individual information users will often be part of several communities. The individuals that comprise a community need not live or work near each other, thus encouraging libraries to address the needs of smaller communities that might have been invisible in the paper era and new communities that did not exist in the paper era. Collections and services for one large Designated Community could also be further customized to meet needs of its own sub-communities. For example, a GIS service that collects and preserves and provides services for geographically based government information might be further customized for new users and for experienced users, or for specific disciplines such as city planning, or agriculture, or water use. The building of collections and services and any customization need not take place at any one library or archive. One institution might build a defined collection, and a second might build a service for a Designated Community using that defined collection. A third library might customize that service for a sub-community. The local communities of all institutions would be able to access and use all collections and services as if they were built locally by their own library.

Such Designated Community collections and services can be designed and built at scale based on two properties of digital information. First, a single digital object can be in many collections by simply adjusting metadata, and multiple copies (stored for preservation redundancy) can all be referenced with the same Persistent Identifier (PID). Second, no matter how course or granular any given digital object is, it will always be connected to its archival provenance and context through the use of metadata. A user might look for and find and use a USGS map, but the user can always find the article or report where they map appeared, the journal or series in which it was published, and all the descriptive information about its creation. The discovery services that the library builds with metadata will enable a user that is looking for a piece of testimony by a specific person to find that testimony easily. But it will do more. A user will be able to seamlessly navigate to the hearing in which the testimony was given and to the bill the hearing was about, and to other hearings on that bill, and to the law that was ultimately passed and signed into law, and to the regulations that resulted from law, and to court cases adjudicating the law. All of this information will be in the comprehensive collection built using the DPI and will be linked by metadata. We refer to this comprehensive collection as the National Digital Archive of Government Information (NDAGI). Each bit of information may reside in multiple defined collections, which may suggest to the user new ways to discover, browse, and search for related information.

In this way, defined collections are no longer information-silos cut off from each other but are integral parts of the comprehensive collection. When specialized collections and services were created in the paper era, they tended to remove information from larger collections and isolate it in a small, focused collection. (This led to "documents departments" in the basements and attics of libraries.) From the user's point of view, finding such information usually requires finding the silo and then figuring out how to use it. Once the user finds the information in the silo, it is rare for it to be connected to related information in other silos. In the digital age, libraries can reverse this model, which was designed for physical objects (books). Instead of organizing discovery and delivery around physical objects for local use, the library can use metadata to organize selected information for communities of users. The defined collections do not remove information from the complete collection but add new ways of discovery and use to that information. That is not just automation or innovation; it is transformation.

In the paper era, the one-to-one model was necessary to ensure that materials would be available and accessible to the library's constituency. In the digital age, the many-to-many model is necessary to make the huge bulk and variety of information discoverable and usable by different communities in different ways (Ammerman, 2020).

The many-to-many library model builds on a familiar approach that many libraries have used in the paper era and transforms it for the digital age. Indeed, the popular movement to share centralized paper collections through better bibliographic control and digitization (Malpas, 2011; Lavoie et al., 2012; Dempsey et al., 2014) is a paper-era implementation of the many-to-many model. At least one project of this movement, the ASERL "Centers of Excellence," began as an attempt to minimize space requirements for housing printed government publications but evolved into a collections and service model (Association of Southeastern Research Libraries, 2012). The ASERL

Centers of Excellence were originally designed to eliminate "excess copies" of printed government publications from ASERL libraries (Burger et al., 2005) and support the management of print collections (Russell, 2010). Each center creates a comprehensive collection of US government information for an agency, or a subject area, or a format (Association of Southeastern Research Libraries, 2018). But the recommended best practice for these physical collections did suggest that they should also have "collection-level expertise for in-depth user assistance" (Association of Southeastern Research Libraries, 2012). In the digital age, libraries can transform themselves by applying the same logic to born-digital information. They can build defined collections and couple them with digital information services for Designated Communities. This would have direct benefits for users, enhancing discovery, access, and usability for many underserved Designated Communities. By making information more easily discoverable, it would result in more information being used more, thus converting the information's potential value into actual value (Chapter 1). Rather than being designed to benefit libraries that wanted to discard their printed government documents, this model is designed to benefit users in transformative ways.

DPI Functionality

The framework allows the development of strategies that combine the functional strengths of archives with the functional strengths of libraries. A distributed implementation of OAIS provides traditional archival functions, and the many-to-many service and collections model provides traditional library functions. In a successful DPI, these functions will complement and overlap each other, thus enhancing the value of both.

A distributed OAIS approach will help ensure comprehensiveness, authenticity, provenance, and integrity. It enables the preservation of the traditional scope of archives (unpublished Records of government) as well as the traditional scope of libraries (published Public Information).

By distributing the responsibility for selection to many libraries and archives, this approach expands the criteria for preservation beyond the limited perspective of the agency that created the information and beyond the limited scope of any single library or archive.

The many-to-many model enables any library to use metadata to define a collection for a specific Designated Community. It enables any library to use metadata to develop digital services for any Defined Collection. This will result in different views of the archival record from different perspectives, thus enhancing the value of the information to users from many diverse communities (Derrida, 1996; Shepherd, 2018). It will also allow libraries to customize the discovery, access, and delivery of information for specific Designated Communities. Different communities will be able to look for, acquire, and use information in ways that best match the way they think about and look for and use information.

The framework allows the development of a DPI in which the parts function together to ensure preservation of all the information needed by people and to enhance services to people. It is guided by information preservation and the information needs of people.

CHAPTER 21
A Distributed OAIS

In this chapter, we outline one possible way a future Digital Preservation Infrastructure (DPI) for government information might implement a distributed OAIS. We describe how to distribute OAIS functions from two perspectives. From the perspective of libraries, we describe how individual libraries could take on specific OAIS roles that match their existing resources and needs. Then we examine the distribution from the perspective of the six OAIS functions. The distribution of functions becomes practical when libraries use the many-to-many model of delivering library services. Thus, this vision of a future DPI incorporates both transformative approaches to preservation we described in Chapter 20.

In planning terminology (Chapter 19), this distribution is a Strategy. It provides a practical strategic-level approach to addressing two immediate problems: the enormous scale of preserving government information and the risk of catastrophic losses. It does this by expanding the scope, capacity, and resilience of archival preservation. It also enhances accessibility and addresses more information-needs of more users. And it makes it practical for more institutions to participate in the government information ecosystem, which will, in turn, result in benefits to the participating institutions and the communities they serve.

Distributing OAIS functions means that, instead of a single institution having to perform all OAIS functions, two or more institutions, operating under a single goal-centered Preservation Plan, collaborate to provide all OAIS functions. (This is not unlike the shared responsibilities for regional depositories libraries [GPO Superintendent Of Documents, 2018].) A distributed OAIS model does not replace the preservation of government information by single-institution implementations of OAIS. Such single-institution implementations can operate either using different preservation plans (with, for example, different scope or goals), or as a collaborative part of the distributed multi-institution plan, or both.

What we are describing here is a vision of a possible solution. This is an exercise in envisioning how the moving parts might work together. The specifics we include in the outline are only examples, not predictions, and are meant to make the vision more tangible and less theoretical.

As a reminder (from Chapter 15), the six OAIS functions are Preservation Planning, Ingest, Data Management, Archival Storage, Administration, and Access. They are described briefly and illustrated in section 4.1 of the OAIS specification (Graphic 21.1).

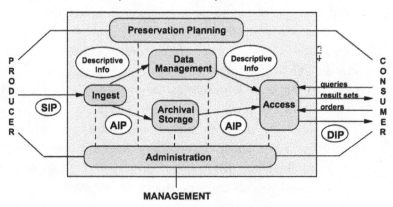

Figure 4-1: OAIS Functional Entities

Graphic 21.1. OAIS Functional Entities (ISO, 2012b).

Libraries as Repositories

In our vision, we distribute the OAIS functions among five different kinds of repositories. Two committees provide direction, standards, and support.

This model gives every library and archive several different options for participating in preserving and providing access to government information. It will thrive by allowing each library to customize its participation based on its resources and the needs of the community it serves. Every different kind of participation provides benefits to the participating library and contributes to building a comprehensively archived national collection. The distribution of functions provides opportunities for every library, no matter how large or small. Its flexibility will also encourage groups or consortia of libraries to collaborate to provide a function. This strategy can start out small with very few participants but can grow without limits.

The five kinds of repositories and the committees are:

- **Ingest/Staging Repositories**. Libraries that target specific agencies, or subjects, or types of content and ingest documents and data for preservation. They process what they ingest and deliver ready-to-preserve information to Primary Archival Repositories for permanent storage and maintenance. There can be an unlimited number of these.

- **Primary Archival Repositories**. A small number of very large repositories that provide permanent storage and management of information. All preserved digital objects are freely and publicly available with PIDs. Each repository preserves a segment of the comprehensive collection. Collectively, they store the comprehensive collection of federal government information— the National Digital Archive of Government Information (NDAGI).

- **Replication Repositories**. Each Replication Repository duplicates the content in one Primary Archival Repository. They use technologies (such as LOCKSS) to detect and repair any damage to preserved digital objects. There can be several Replication Repositories for each Primary Archival Repository. Some can expose their copies with the same PIDs used by Primary Archival Repositories, while others are "dark." These are inexpensive to run and there can be dozens of them.

- **Defined Collections and Services Repositories**. Libraries will define collections using metadata and provide digital services (discovery, delivery, use) to a Designated Community. They need not have their own copies of digital objects, but they can if they wish. (Some libraries may wish to retain copies to index or repackage information.) There can be many of these, and they can vary in size. A library can host more than one Defined Collection or Service. This model is similar in concept to the Association of South East Research Library (ASERL)'s "Centers of Excellence" model in their Collaborative Federal Depository Program (CFDP).

- **Metadata Repositories**. One or two repositories that preserve metadata for the complete archive. They also describe what is available for preservation and what has been preserved. Replication Repositories ensure their integrity.
- **The Preservation Planning Committee**. Defines the Goals and Strategies of the Digital Preservation Infrastructure.
- **The Preservation Implementation Committee**. Provides technical standards and procedures for the day-to-day implementation of the objectives, tactics, and tools of the Digital Preservation Infrastructure.

The OAIS Functions

Preservation Planning

The OAIS Preservation Planning function provides recommendations and preservation plans to ensure that the information stored in the OAIS remains accessible to and understandable by the Designated Communities over the long term, even if the original computing environment becomes obsolete.

Preservation planning is the key to a successfully distributed OAIS. Using a goal-focused hierarchical planning process (described in Chapter 19), it will provide the standards and methods that will support actions toward a common goal. This will enable separately functioning projects to interoperate.

Preservation Planning is centralized in a Preservation Planning Committee consisting of librarians and archivists from government agencies and from non-government libraries and archives. Its responsibility is to define the goal, design strategies, choose standards, and create policies for the administration of objectives and tactics and use of tools.

A separate Preservation Implementation Committee is responsible for the day-to-day administration of the strategies defined by the Planning Committee.

By defining the goal and developing the strategies of a Preservation Plan, the Preservation Planning Committee would ensure that the objectives, tactics, and tools of all distributed projects and functions conform to the goal of preserving government Public Information.

Ingest

The Ingest function includes accepting Submission Information Packages (SIPs) from producers and preparing the contents for storage and management within the archive.

The most fundamental of the gaps we identified in Part Two is the preservation gap. This gap indicates the failure of the existing infrastructure to ingest information that deserves preservation. To bridge that gap, it is essential to increase the capacity of the infrastructure as a whole to identify, select, and ingest information efficiently and effectively.

There are two primary challenges to increasing the capacity of ingest. The first is the quantity of information under consideration. There is a lot of content on the government web from which to select, and it changes and moves and even disappears. Finding information that is new or that has changed and identifying it and determining if it matches specified criteria for preservation can be a complex task that takes a lot of resources. Doing so before it moves, or changes again, or is withdrawn entirely requires agility and speed.

The second challenge is that there are many kinds of information on the web (e.g., text, numeric data, images, A/V) that come in many different file formats. Each different kind of information requires specialized attention for ingesting it properly, for collecting or creating metadata for it, and for storing it in appropriate Archival Information Packages (AIPs) so that it can be delivered in appropriate Dissemination Information Packages (DIPs). In some cases, the same information is available in different formats and presentations and packages. Decisions need to be made about which presentations (sometimes more than one) need to be preserved based on the way different Designated Communities will use the information.

The old way of dealing with these challenges is to use a generic approach, such as web harvesting, that treats all information the same way—as "content" on web sites. Currently, the use of large-scale, one-size-fits-all, web-harvesting mega-projects is the primary method of trying to save digital government information published on the web. But, as shown in Part Two, that approach can miss useful content, unnecessarily ingest multiple copies of the same content, and ingest content based on how easy it is to harvest rather than on its intrinsic value. Information that cannot easily be captured through conventional web archiving practices is "endangered" according to the Digital Preservation Coalition (Digital Preservation Coalition, 2023).

We suggest that a distributed implementation of OAIS will create a more efficient and effective approach by using focused ingesting streams. Each ingest stream will be customized for a particular source or kind of information. Using ingest streams is not radical or even new; it is already in common use to a limited extent. The Library of Congress, for example, uses this approach with its web harvesting, which focuses on individual subjects or themes or events and presents these to the public as "collections" (US Library of Congress, 2025a; US Library of Congress, 2023). GPO also uses focused web harvesting with its FDLP Web

Archive program, which focuses on small websites (GPO, 2021a; Bower & Walls, 2014). Focused ingest is not limited to web harvesting. A large part of what GPO ingests into GOVINFO comes through 38 ingest streams that GPO refers to as "Collections." Each of those streams consists of content that has a consistent format such as "Congressional Hearings" (GPO, 2021e).

Focused ingest is not magic, though. It still requires a lot of work, but it defines and focuses that work. Even narrowly defined content types require careful attention at ingest. Consider, for example, academic journal articles as a well-understood type. Many publishers produce articles in well-structured, well-defined file types such as the NLM Archiving and Interchange Tag Set (US National Library of Medicine, 2012), but even this well-defined file type comes in more than 100 varieties and journal articles are often accompanied by supplemental materials, which can come in any of more than 100 different file formats (Morrissey et al., 2010). A Center for Research Libraries preservation audit of the Portico e-journal archive reported that Portico has to transform publisher-supplied content into Submission Information Packages (SIPs) (Center for Research Libraries, 2010). Still, an ingest stream for one kind of information such as this can more easily achieve efficiency and effectiveness than an ingest stream for all kinds of information lumped together. Focused ingest does not eliminate complexity, but it narrows the amount of complexity needed for any given ingest stream. It is a practical and manageable way to cope with the challenges of the variety of published Public Information.

A successful DPI must support selecting and ingesting all the digital objects that librarians and archivists deem worthy of preservation based on the information-needs of the communities they serve. It can do this by supporting procedures and standards that make it easy for libraries to create ingest streams. This will increase the number of ingest streams and distribute the OAIS ingest responsibility across many institutions—all under a common, goal-focused Preservation Plan. Each Ingest Repository chooses to take the responsibility for one or more focused ingest streams.

This will increase the capacity of the DPI as a whole to cope with many more sources and kinds of information and will do so more efficiently and effectively than any single repository could.

This will involve both large-scale, comprehensive, provenance-based ingest and smaller scale subject or data-type ingest. This will enable more institutions to participate in selecting and ingesting data. As more institutions participate, more content will be ingested and less content will be missed.

Each Ingest Repository will choose its focus using its own specific criteria. Using technical criteria, a repository can focus by kind of information (e.g., text, A/V, numeric data, dynamic database), or file type (e.g., PDF, GIS, database), or intended use (e.g., close reading, distant reading, statistical analysis). Using provenance, a repository can focus on a branch or agency or office of government or on a web domain or web server or website. Using information content as criteria, a repository can focus on content by discipline (economics, agriculture, law) or subject (water quality, consumer affairs), or region, or event, or person, and so forth. Because ingest streams define content as belonging to a type or subject or source, these definitions will also be used to enhance accessibility of content for users. Libraries and archives will be able to define browsable collections and develop enhanced discovery tools.

Each Ingest Repository will choose and define the criteria for ingest based on several parameters. These include: the needs of its own Designated Communities, a publicly accessible record (in the Metadata Repository) of what other ingest streams already exist and what they have ingested, and explicit or implied cooperative agreements with other libraries and archives.

The DPI will support all sizes of Ingest Repositories. Institutions that can support a large-scale Ingest Repository will be able to choose to ingest comprehensively, selecting content from an agency or a branch or a government web domain. Other institutions will be able to choose an appropriately sized focus such as a subject or type of data. Smaller institutions will be able to select content on a small, specialized scale.

For example, a library might select everything it can find by and about local dignitaries such as Representatives and Senators, or about an issue that is important locally such as water management or agriculture. This model expands the scope of preservation by focusing on information rather than on websites and by actively ingesting wanted information that general-purpose web harvests might have missed. It also empowers libraries to ensure that the specific data-types of content ingested match the use-needs of specific Designated Communities.

The work of selection and ingest itself can be split among different institutions by categories of tasks such as identification of new content, selection of content for ingest, acquisition of content, preparing content for ingest, and ingesting and creating metadata. This division of labor will allow even small institutions with limited resources to identify and choose an appropriate scope of participation and provide significant contributions to the large-scale distributed OAIS project.

We anticipate that identification of new content for potential ingest will become a new information preservation activity and an essential element of the DPI. The results of scans of the government web will be stored in a Metadata Repository. Ingest Repositories will be able to use this information to target content for ingest.

Ingest Repositories are temporary Staging Repositories (described in Chapter 15). After acquiring and processing the information it targets for preservation, an Ingest Repository delivers it to an appropriate Primary Archival Repository for permanent archival storage and can then delete its copy. Upon being stored in a Primary Archival Repository, each digital object is assigned a Persistent Identifier (PID).

Using standards specified by the Preservation Plan, each Ingest Repository will share its focus and results with the other Ingest Repositories. This information will be stored in a Metadata Repository, which will be the publicly accessible record of what is being ingested. By sharing their targets of preservation, what they find, and what they ingest, the distributed ingest system maximizes coverage and comprehensiveness while reducing unnecessary duplication. Inevitably, many digital objects will fit the criteria for multiple ingest streams. The Preservation Implementation Committee will coordinate ingest priorities by specifying and facilitating the use of metadata among the Ingest Repositories to ensure that objects are ingested only once. Objects can be tagged in metadata as belonging to more than one ingest stream. The Metadata Repository makes it possible to identify, characterize, and quantify information published and information preserved. This will help bridge the preservation-information gap identified in Chapter 13, making it possible to analyze gaps in preservation.

Government agencies could easily act as ingest repositories for their unpublished Records, or their published Public Information, or both by following the guidelines, requirements, and standards specified by the Preservation Plan. NARA could encourage this with guidelines and requirements that follow the Preservation Plan's strategies and standards.

Since many ingest streams are treatable as collections, some Ingest Repositories may choose to keep a copy of the data they ingest for further processing in order to develop advanced tools for discovery, delivery, and use.

Some Ingest Repositories will ingest little or no data but simply enhance the metadata of already ingested digital objects by tagging them as matching a particular ingest-category and then ingest just the data missed by other ingest streams.

Archival Storage

Archival storage provides the services and functions for the storage, maintenance, and retrieval of AIPs.

Archival storage is the foundation around which all the other functions revolve. If the DPI can guarantee that selected content can be ingested and stored so that it will be permanently available at a persistent address, then it will encourage selection and ingest and provide the foundation for the development of digital services.

Because there is so much Public Information to store and because it comes in so many different forms, it will be necessary to have more than one or two Primary Archival Repositories. Therefore, a successful DPI will enable the distribution of responsibilities into segments among Primary Archival Repositories with the guidance of the Planning Committee. For example, if there were just two Primary Archival Repositories, one at GPO and one at NARA, digital objects might be distributed between the two with NARA preserving unpublished Records and GPO preserving published Public Information. This would be much like the paper-era division of responsibility between the two agencies. Thomas Brown, who managed NARA's Archival Service Electronic and Special Media Records Services Division, suggested that such a division of responsibilities would be an appropriate way of implementing T.R. Schellenberg's principles of archival management in the age of web records (Brown, 2006).

Much of the archival storage infrastructure of a future Digital Preservation Infrastructure already exists at GPO, NARA, and the Library of Congress, which are already preserving government information in bulk. These agencies, which also already have a legal mandate to preserve government information, can create the backbone of a system by assuming the role of the first Primary Archival Repositories.

In practice, however, there will likely be more than three Primary Archival Repositories. Any agencies that have Congressionally mandated and funded repositories that meet the requirements of the Preservation Plan will be eligible to participate as Primary Archival Repositories. The division of responsibilities among Primary Archival Repositories will be defined by criteria such as branch of government or type of information or other factors as determined by laws, regulations, and the Preservation Plan. Designation as a Primary Archival Repository will be defined as accepting the Preservation Plan and following its requirements.

At least one and possibly several Primary Archival Repositories will reside outside of the federal government. Such non-government Archival Repositories will be useful for preserving government information that falls outside of the legal remit of all of the government's Primary Archival Repositories. That might include, for example, government information posted on commercial social media websites, and data and publications created with government funding. When laws limit the authority of government agencies to preserve information that that Ingest Repositories identify as worthy of preservation, non-government Primary Archival Repositories will provide a permanent home for such information. Non-government Primary Archival Repositories will ensure the preservation of government information whenever the government abrogates its responsibility for preservation for any reason.

Each preserved digital object will be stored in only one Primary Archival Repository. The Preservation Plan and its implementation through segmenting archival storage responsibility will minimize unnecessary duplication of data stored in Primary Archival Repositories. Although Primary Archival Repositories will have their own duplication and backup services, the Preservation Plan will mandate a strategy of using Replication Repositories outside the government to protect integrity and to ensure access. Each Replication Repository will automatically receive a copy of all data stored by one Primary Archival Repository. (This functionality is already in place at 36 libraries with the Digital Federal Depository Library Program, which is also known as the LOCKSS USDOCS network, which protects information stored in GPO's GOVINFO repository [LOCKSS Program, 2022].) Replication Repositories will use integrity monitoring and repair technologies similar to that of Chronopolis (University of California San Diego Library, 2023) or LOCKSS (Maniatis et al., 2005) to protect against loss or corruption of the copies stored in Primary and Replication Repositories.

There will be several Replication Repositories for each Primary Archival Repository in order to ensure adequate integrity checking. Each Replication Repository is either "light" (data available to the public) or "dark" (data not yet available to the public) (Reilly, 2008; Shah & Gul, 2019). The data in both the light and dark Replication Repositories is used to identify and then replace or repair lost or damaged data. As mentioned in the previous chapter, the use of Replication Repositories also prevents any intentional political decision to alter or remove information from public access once it has been archived. The purpose of the DPI is to preserve government information, not to be an instrument of shifting government policies.

As the size of the complete archive expands, more libraries and archives will be able to choose to become Replication Repositories. The costs (both hardware and staffing) of providing dark replication repository services will be kept very low to encourage participation and, by relying on PIDs, the cost of providing light replication repository services will be only marginally higher.

The data in publicly available ("light") Replication Repositories will be available using the same PIDs assigned and used by the Primary Archival Repositories. This requires that the PID technology used is capable of multiple resolutions—a single PID that stores the addresses of several copies and either automatically delivers a copy or allows the user to choose among copies. (The Digital Object Identifier [DOI] technology already has this functionality [DOI Foundation, 2023].) Multiple resolution to several copies stored with different institutions ensures uninterrupted availability of all data even in the event of government shutdowns or the loss of information caused by inadequate funding, political decisions, technical malfunctions, changing government priorities, or other interruptions of access to government-run repositories. Multiple resolution PIDs will also be able to point to the original agency copy of information and any subsequent agency addresses as well as the permanently preserved copies. This will facilitate creating services that direct users to preserved copies of information when they only have link-rotted URLs. Multiple resolution PIDs have the additional long-term benefit of enabling future technologies such as having a single PID provide access to different versions of the same content (i.e., data types [Philipson, 2019] or editions [Gleim & Decker, 2020]).

Altogether, these actions implementing the Archival Storage function of OAIS will preserve a comprehensive National Digital Archive of Government Information.

Access

The OAIS Access Function provides services that support consumers in determining the existence, description, location and availability of information stored in the OAIS, and allowing consumers to request and receive usable information products. This functionality is distributed between Primary Archival Repositories, light Replication Repositories, and Collection Services.

Availability. Every digital object will be preserved in the National Digital Archive of Government Information (NDAGI) and retrievable using its PID. Each PID will, at minimum, point to a copy in a Primary Archival Repository and to all the copies in the (light) Replication Repositories, thus ensuring uninterrupted availability of all data.

Accessibility. The Preservation Plan will define a basic level of accessibility for all Primary Archival Repositories, including searching and browsing by traditional descriptive facets such as author, title, subject, date, defined collection, ingest stream, and type of information (e.g., file type). Primary Archival Repositories would also provide at least a basic method for delivering bulk access to their data. Primary Archival Repositories and Metadata Repositories will also expose their contents and metadata to web search engines for public indexing. The Preservation Plan will require that, at minimum, every ingested digital object is deliverable to users as part of at least one Dissemination Information Package (DIP).

This will satisfy the bare minimum of the requirements for preservation. It ensures that information is safely saved, protected from alteration or loss, and that it and its descriptive information are made available with a permanent address. But preservation requires more than availability; saved information must also be strategically accessible to Designated Communities.

Saving information is largely about accommodating the needs of the information itself. Making information accessible is largely about accommodating the needs of users. The two are inextricably related but very different.

Thus, full preservation services will require specialized accessibility beyond the basic availability provided by Primary Archival Repositories. Collection Services will provide this accessibility.

Collection Services will be designed, created, and maintained by libraries and archives that wish to enhance the accessibility and usability of government information beyond the basic level provided by Primary Archival Repositories.

Each Collection Service will define and select content for a collection of government information and develop digital services for discovery, access, delivery, and use of that Defined Collection based on the needs of a Designated Community. For example, a library might describe a collection of government information pertaining to the economy, gathering information from many agencies. Another library might build a service this collection aimed just at economists and anther library might build a service on this same collection aimed at the general public.

Existing library consortia will be able to take responsibility jointly for shared interests and services. Discovery and delivery services for each collection will be designed to match the needs of users of that collection. Discovery services will include interactive web interfaces and APIs (Application Programming Interfaces). Delivery services may offer customized DIPs including bulk data delivery as appropriate to the collection.

This approach will directly benefit users and individual libraries. Each library will likely choose collections and services based on the needs of their immediate constituencies but will make its Collection Services available universally. All users everywhere will have access to all Collection Services. In this way, individual libraries will be able to provide specialized collections and services for many of their different constituent communities without having to develop all those collections and services themselves. Library users will benefit by having access to Collection Services developed nationally, not just to those few that their own library can afford to develop.

Government agencies will also be able to provide Collection Services by conforming to the Preservation Plan. Some agencies might provide more than one OAIS function. For example, congress.gov might serve as an Ingest Repository, a Primary Archival Repository, and a Collection Service. It would do so by adopting the standards of the Preservation Plan and referring to preserved content by their assigned PIDs.

Some Collection Services will choose to host a copy of the content in their collection in order to develop unique discovery and delivery services. In such cases, their copies are added to the multiple resolution locations for the originally assigned PIDs. Some Collection Services may create new AIPs (Archival Information Packages) or DIPs (Dissemination Information Packages) in order to meet the needs of a specific Designated Community. For example, a library might want to make information available at a more granular level (e.g., articles in a serial, variables in a database) than is available from the Primary Archival Repository. It might want to have local copies to facilitate the development of full text indexes or other discovery services. It might want to be able to deliver customized DIPs that bundle disparate items to match user specifications. When a Collection Service creates new, permanent AIPs or DIPs, it will offer them to an appropriate Primary Archival Repository for archival preservation and additional availability. New, permanent DIPs will be assigned PIDs.

The Preservation Planning and Implementation Committees will provide standards for constructing URLs that refer to temporary or dynamic DIPs (such as a table from a database). That standard will define the construction of a temporary item's URL that reconstructs the content of the temporary or dynamic DIP and identifies its provenance (AIPs).

Collection Services will use metadata to define their collections and to provide discovery services. They will be able to draw on existing metadata in the Metadata Repositories, and they will be able to create their own metadata, which they can offer to the Metadata Repository for others to use.

Since ingest streams may be usable as collections, metadata would identify all the content of each ingest stream and make them available for the development of Collection Services.

Administration

The OAIS administration function provides services and functions for the overall operation of the Archive system. Generally applicable administrative functions are centralized in a Preservation Implementation Committee. It manages the implementation of the Preservation Plan and administers generally applicable administrative functions by providing the services and functions for the operation of the OAIS as a whole. The Implementation Committee ensures that the strategies developed by the Planning Committee are implemented.

Individual repositories and projects administer their own day-to-day management of their operations using guidelines and standards defined in the Preservation Plan.

Data Management

The OAIS data management function provides services and functions for populating, maintaining, and accessing Descriptive Information, which identifies and documents Archive holdings, and administrative data used to manage the Archive. Each individual Repository ingests, creates, maintains, and updates metadata it needs to manage its data and does so using the standards and guidelines defined by the Preservation Plan.

Metadata is data, and one or more centralized Metadata Repositories manage and preserve metadata for the system as a whole. Metadata is integral to the management and preservation of data and to the discovery, delivery, and use of the data. PIDs are a form of metadata, and assignment and maintenance of PIDs will be a large and essential role of one or more Metadata Repositories. Metadata must be actively and immediately available to users in order to implement dynamic discovery and delivery of data.

Metadata is, itself, a valuable kind of data. The Metadata Repositories will provide user services that will make metadata available to users for discovery, analysis, and use. The DPI will also provide a way for users to contribute metadata and to identify inaccuracies in metadata.

Conclusions

The DPI that we describe here will address the gaps that we identified in Part Two. Specifically:

1. The inventory gap. The Metadata Repository will contain information both on the production of government information and on its preservation. This data will be publicly available for examination and analysis, enabling those interested in preservation to have a more accurate and more complete inventory of born-digital federal government Public Information and the extent of its preservation.

2. The preservation gap. By creating more ingest streams, the DPI will be able to ingest more information and ensure that less information is missed for preservation. By guaranteeing permanent storage and management of ingested data, the DPI will preserve a comprehensive National Digital Archive of Government Information.

3. The legal gap. The DPI will work within the bounds of existing laws but will expand the reach of preservation beyond the limitations of existing laws. The DPI will be a preservation infrastructure, not an instrument of government policy. From the perspective of the DPI, government agencies are free to

participate. The ongoing planning function will define and advocate for changes in laws and policies to better support the DPI, including agency participation, as it develops and evolves.

4.   The infrastructure gap. The DPI will facilitate preservation of government information by providing a culture of preservation and an "infrastructure" as described by Star and Ruhleder. All libraries and archives will be able to participate easily. It will address accessibility of preserved information for Designated Communities. Its success will build on the traditional strengths of libraries and archives: selecting, acquiring, organizing, and preserving information and providing services for that information to their communities. But it will change how libraries accomplish this, converting to a distributed OAIS model and a shared, many-to-many approach to services that will reach more communities with more services for more information.

5.   The preservation-information gap. With standards promulgated by the Planning Committee and used by the Preservation Repositories, the DPI will provide more consistent and more accurate information about what is preserved. This information will be preserved in the Metadata Repository and made available to all for exploration and analysis.

6.   The preservation planning gap. The Preservation Planning Committee will provide a goal-driven planning process usable across branches of government and by non-government libraries and archives. The existence of a plan for digital preservation will provide a context for government agencies to create preservable digital information.

7.   The gap in responsibility (Chapter 18). Library and archive administrators and managers will be able to demonstrate to their constituents that they are meeting both the immediate and long-term needs of their institutional community as a whole as well as many of its sub-communities. Administrators will be able to demonstrate that they are doing this by participating in a

national, goal-driven preservation plan that integrates collections and services with preservation.

CHAPTER 22
Many Roles

In the previous chapter, we took a top-down approach, focusing on the OAIS responsibilities and how they could be distributed across many institutions. In this chapter, we take a bottom-up approach, shifting the focus to the institutions (governmental and non-governmental) that will actually do the work of building a National Digital Archive of Government Information. In the context of a new Digital Preservation Infrastructure, each institution will have the opportunity to choose among a wide variety of roles. Each one will be able to participate in preservation without having to assume every preservation responsibility. Similarly, each library will still choose how to use its own resources to serve its own communities, but each will gain the benefits of access to the collections and services developed by all libraries.

Government Roles

It is clear from the many "principles" of government information described in Chapter 1 that there is a broad consensus that the government has an obligation in the preservation of its own information. In the paper era, that obligation was addressed by publishing and distributing Public Information and relying on libraries to preserve and provide access and services for that information. In a DPI in which OAIS responsibilities are distributed across many institutions, which responsibilities are the ones that the government should, or must, assume?

Digital preservation is not a product but a process, and the process begins when the digital object is created. As the producers of the information to be preserved, government agencies are in a unique position to affect digital preservation. This is where the government obligation drives its responsibilities.

Publishing Role

One essential, but less-discussed, government role in preserving its own information is its creation and dissemination of Public Information. In order for the government to honor its obligation to preserve its information, agencies must recognize that obligation and make their information easy to preserve. This begins with basic communication. Agencies must make it easy for preservationists to identify new information when it is published—or altered. Secondly, they must make it easy for preservationists to acquire sufficient control over the information to preserve it. Technologies for announcing and listing and describing and distributing information exist. If the government takes its obligation seriously, it will standardize and enforce the use of such technologies by agencies that publish Public Information in order to ensure its preservation. In the digital age, this is an essential preservation role, and the government itself is uniquely placed to fulfill it.

Agencies that produce digital Public Information should also be required to create digital objects that contain Preservable Digital Information. Government agencies, like all of us, adopted digital publishing to the web using the tools available. Unfortunately, many, if not most, of the tools that technologists developed emphasized speed and ease of production, not preservation. This means that, in many cases, the digital objects that deliver Public Information make it difficult to preserve the information they contain. This trend should be replaced with a publication infrastructure that produces, by default, Preservable Digital Information. This is not a simple task but, until it is addressed, government publishers will continue to make the task of preservation more complex and difficult and expensive than it needs to be. The scale of government publishing provides government with an opportunity to play a major role in influencing the development of standards for

Preservable Digital Information. The role of producing Public Information that is easy to preserve is another role that is essential and for which the government itself is uniquely placed to provide leadership. (See also Appendix D about the "significant properties" of digital information.)

Planning Role

Planning, as described in Chapter 19, is an essential component of a successful DPI. The government has an opportunity to play an exceptionally large part in planning for the preservation of government information because of its unique position as producer of the information to be preserved. It will, by design of the DPI, participate in the Planning Committee and its work on the development of Goals and Strategies.

There are, however, many additional planning roles that only the government can assume. In its role of developing and enforcing laws, policies, regulations, and procedures, it must ensure that digital preservation becomes an explicit priority. Government information policy is extensive (see Chapter 5), but these policies have tended to neglect preservation more than attend to it, and, when they do address preservation, they have treated it more as an afterthought or an add-on to print publishing rather than as a priority. Government information policies have a direct effect on preservation, so a unique role for government is to prioritize information policies that ensure preservation. Having a well-designed, goal-based preservation plan developed by the Preservation Planning Committee will provide agencies a framework within which it will be easier to develop policies that are practical and effective. The information policies that affect preservation include those that affect the discovery, description, preservability, and distribution of digital information. If this planning is successful, it can make the job of preservationists easier and less expensive; if it is omitted or done badly, it will make it difficult for preservationists to identify and select and acquire content for preservation.

Archival Storage Role

In the DPI that we envision, the permanent storing of digital objects and making them available with Persistent Identifiers is the foundation upon which digital services for access and delivery and use of those objects can be built. The government is in a logical place in the information ecosystem to fulfill this foundational role by providing Primary Archival Repositories. While this is a very large job, it is a practical one for government to assume. The distribution of OAIS responsibilities makes it possible for agencies to take on large tasks such as this that focus on a single responsibility, rather than having to assume all OAIS responsibilities.

Splitting the responsibility for archival storage among several Primary Archival Repositories will make it easier for agencies to participate by scaling their participation to an appropriate and sustainable level. It would be easiest for traditional preservation agencies such as GPO, NARA, and the Library of Congress to be the first participants as Primary Archival Repositories. But agencies that produce information could also host their own Primary Archival Repositories if they had a legislative mandate and funding to do so. (Some self hosting already exists with minimal or no legislative preservation mandate. Examples include repositories at the national libraries of medicine, education, agriculture, the Smithsonian, the National Oceanic and Atmospheric Administration, the Defense Technical Information Center, the National Technical Reports Library, and NASA's Technical Reports Server.) Each participating agency will be able to focus on a defined body of data to preserve. This will reduce the quantity of data any given repository will have to handle. If, in choosing to participate as a Primary Archival Repository, a publishing agency proves to be a reliable selector of published Public Information for preservation, it will also reduce the load of all other Ingest Repositories.

Splitting archival storage responsibility will also remove the difficult burden of any single repository having to cater to all types and sizes of data. This will allow Primary Archival Repositories to specialize in, for example, the way it stores and delivers a particular type of data. This permits repositories to avoid either providing only cumbersome, one-size-fits-all, lowest-common-denominator ways of storing, managing, and delivering data, or trying to support multiple specialized ways of dealing with different kinds of data. This will reduce the overhead of the archival storage tasks for individual repositories while simultaneously providing better preservation and better user services in aggregate.

Primary Archival Repositories will also benefit from the existence of non-government Replication Repositories. The Replication Repositories will increase the overall security and resilience of the OAIS system as a whole, lessening the burden on the Primary Archival Repositories. By spreading access copies to institutions outside the government, this approach also protects against interruptions of service caused—intentionally or unintentionally—by government shutdowns, government policies, actions, or funding shortfalls.

Metadata Role

A successful DPI will rely heavily on accurate, extensive, flexible metadata. Libraries that provide collections and services will use metadata to define their collections and maintain the connection of the information in those collections to their context in the comprehensive National Digital Archive of Government Information. They will also use metadata to design discovery tools for users. They will use metadata to structure and define information objects suitable for delivery as Dissemination Information Packages (DIPs).

Government agencies that produce Public Information are in the best position to accurately describe the information they produce. Their role in producing metadata is not unique, but it is primary and essential. As such, along with its publishing role, government agencies are in the prime position to assume the role of creating essential, definitive descriptive metadata for the information they create. Requiring agencies to assume this role has been tried before without success. (See Appendix E for the example of the *Government Information Locator Service*). But if government is to take its obligation to preserve its own digital information, it must create and enforce the policies and standards that make it possible and even mandatory that the publication of Public Information includes essential, accurate metadata.

Finally, just as government agencies will be perfectly positioned to manage Primary Archival Repositories, government will also be ideally positioned to establish and run one or more of the Metadata Repositories.

Other Roles

Government agencies are not limited to their unique roles; they can also participate in any of the other distributed OAIS roles. As mentioned in Chapter 21, agencies may choose to provide defined collections and specialized digital services. The Library of Congress, for example, already does this with its web-harvesting collections and with congress.gov. Agencies that produce Public Information may choose to serve as Ingest Repositories, thus simplifying the preservation of their own published Public Information and unpublished Records. Some information-producing agencies may choose to partner with specific Ingest Repositories to provide customized delivery of their Public Information for preservation.

Roles for Non-Government Libraries and Archives

Planning

Moving to a distributed implementation of OAIS and a many-to-many model of service will take some profound changes in the social norms of libraries. It will also take careful planning that accommodates lots of different kinds and sizes of institutions. These include government agencies that produce information and libraries and archives that preserve information and provide digital services for defined collections.

A Preservation Planning Committee that includes librarians and archivists from government agencies and from non-government libraries and archives will be essential to this planning process. One of the key responsibilities of the committee will be to generate opportunities for institutions, organizations and individuals to participate in shaping and developing the new DPI. Although a practically-sized committee will only afford opportunities for a limited number of individuals, the work of the committee will inevitably include gathering information from communities of stakeholders. One of the roles that the library and archival communities will have is the opportunity to articulate their own visions and preferences in the planning process as stakeholders. Organizations and individuals will be able to participate in the design of the strategies of preservation and access.

In order to establish and maintain new social norms for the roles of libraries, it will be essential for libraries to understand the needs of their constituents and to be able to express a common understanding of how libraries can meet those needs in the digital age. The committee and its job of articulating such understandings as a Plan with Strategies will be essential. Oya Rieger, in her Ithaka S+R study noted that the managers of libraries and archives need a common understanding of the roles of their institutions, and they need "overall strategies" and "a cohesive and compelling roadmap" for preservation (Rieger, 2018). The Planning Committee will develop the language, standards, and strategies that accomplish this. Membership on the committee will be a vital role for some individual librarians, but the library and archival communities have an equally vital role to play in contributing to the planning process and the development of the plan.

The existence of a carefully designed, goal-driven preservation plan will facilitate the communication between preservationists and their library managers and between libraries and archives and their user communities and funders. It will give a context and vision to the individual roles that libraries assume and the tasks they employ.

Selection and Ingest

In order to preserve more information than is currently being preserved, more people and more institutions need to be actively involved in identifying, appraising, and selecting content. In order to ensure that the right information in appropriate formats is preserved, more institutions need to be involved in selection and ingest than is currently the case. These institutions can reflect the use-needs of more Designated Communities. The current infrastructure of relying almost exclusively on three government agencies (NARA, GPO, and LC) just has not scaled to meet the quantity of digital information produced by the government. But this is not only an issue of quantity, it is also an issue of perspective. A government agency has its own perspective on which information is valuable, but users often have a different perspective and ingesting libraries can express that perspective for their communities.

Involving non-government actors to select information for ingest into Primary Archival Repositories for preservation will require guidance from the Planning Committee. Once the committee has put standards and procedures in place, it will be easy for libraries of any size to find a level of participation that matches the needs of their communities and their resources. With several specialized primary repositories, including at least one run outside of the government, there will be clear, legal paths for deposit of all the content that the Ingest Repositories select for preservation.

Selection of content is so important and so potentially complex that we have included in Appendix C descriptions of some of the issues that need to be addressed.

Below, we describe the sub-roles available to libraries for participating in selection and ingest. Individual libraries may choose to assume one, or some, or all of these sub-roles. The Preservation

Implementation Committee will coordinate availability and assumption of roles.

Identifying Content for Appraisal

In order to appraise and select content, it is necessary to know what content exists. This will require new software tools that are specialized for scanning the government web to identify new and changed content and characterize it by type, content, and subject. A key role for one or more libraries will be the development and maintenance of such software. Different tools may be developed to scan for different kinds of content and different kinds of websites. Once tools are available, other libraries will be able to assume the role of using these tools to scan for subjects, or data types, or to scan specific segments of the government web. With several libraries scanning different portions of the government web, coverage can be maximized while minimizing the cost to individual participating libraries. The results of scans will be posted into the Metadata Repository to provide Ingest Repositories with the information they need for focusing their selection and acquisitions.

Acquiring

Once content has been selected for ingest, another role will be to actually acquire the selected content. Although web harvesting is the primary way information is acquired for preservation today, under the new DPI, acquisition will also be accomplished by using additional techniques and tools that match the nature of the content, its availability, and special curation agreements made with agencies that produce the information. Some acquired content will require modification to meet the SIP requirements of the appropriate Primary Archival Repository.

Metadata Management

Metadata will drive selection, preservation, and access, so the acquisition, creation, preservation, and maintenance of metadata are key to a successful DPI. Key metadata tasks will include: creation and maintenance of PIDs, the automated acquisition of existing metadata, transformation of acquired metadata into standardized formats, and automated and manual creation of metadata. Once metadata is created, its preservation and maintenance (including development of mechanisms for access and delivery of it) will provide yet another role for libraries.

Ingest

Once information is selected, acquired, and described in metadata, the final task of this OAIS responsibility will be the actual transfer of prepared content and metadata for ingest into an appropriate Primary Archival Repository.

Archival Storage

There will be a limited number of Primary Archival Repositories, and many of these (but not all) will be run by government agencies. At least one and possibly several non-government institutions may choose the role of being a Primary Archival Repository. A successful DPI will require non-government Primary Archival Repositories for two reasons. First, at least one will be necessary for preserving content that no government repository can host for legal, policy, technical, or economic reasons. This might include, for example, Public Information that agencies post to commercial hosts such as social media sites. Second, although it would be ideal for the federal government to take primary responsibility for hosting its own preservation repositories, until it takes explicit action to do so, non-government Primary Archival Repositories will be needed for a DPI to be successful. Without laws that explicitly mandate the permanent preservation of all government Public Information, and without solid, bi-partisan, consistent support for adequate funding of preservation repositories, the success of even a well-planned, goal-driven DPI will rest *outside* of government. The purpose of the DPI is to preserve government information. Government agencies can and should participate in that, but the DPI does not require them to do so. The DPI is an open culture of preservation, not a government project.

Hosting a Primary Archival Repository will, in most cases, be a resource-intensive commitment, but we already have models of such commitments within the government (NARA, GPO, LC, NASA's National Satellite Land Remote Sensing Data Archive) and outside the government (the University of Michigan's HathiTrust and ICPSR, the University of Texas' CyberCemetery, the Internet Archive's Democracy's Library [Kahle, 2022]) and many more. These activities exist without a national digital preservation plan and without a DPI. We anticipate that the development of a DPI will attract more institutions because the DPI will provide a predictable and efficient path to demonstrable success. Designation as a Primary Archival Repository will be based on an institution's agreement to the Preservation Plan and its standards and procedures.

There will also need to be many Replication Repositories which will protect the integrity and ensure access to content stored in Primary Archival Repositories. The cost of creating and maintaining these Replication Repositories will be kept very low, and most activities will be automated. This role can benefit from and expand on the 16 years of experience of the USDOCS Private LOCKSS Network, which has been serving this role for the GOVINFO repository (Petrich & Becker, 2024).

Collection Services

As described in Chapter 20, one of the main transformations facilitated by the DPI will be the enabling of digital services for defined collections aimed to meet the needs of specific Designated Communities. Any library that wants to define a collection and develop digital services for it will be able to do so using the metadata in the Metadata Repositories and the data stored in primary and replication repositories. Thus, the first two-thirds of their work will have already been done by others. The Primary Archival Repositories will comprise the comprehensive National Digital Archive of Government Information from which sub-collections can be defined with metadata. Those libraries that need their own copies of the data in their defined collection (e.g., for specialized indexing, repackaging for delivery, etc.) will be able to acquire it in bulk from Primary Archival Repositories.

Each Defined Collection will be a subset of all preserved government information. A defined collection is defined by the specification of criteria that can be matched by content's metadata. Collections can be defined, for example, by subject, source, or datatype. Digital Services for a defined collection might include specialized ways of searching and browsing content, and specialized ways of presenting and delivering content, including packaging it for bulk data delivery. Such services can be developed using existing metadata or metadata created by the service, including metadata created based on the full text of the targeted content.

Some libraries may take on the role of designing software, databases, and digital service modules that can be adapted for use for different defined collections and services.

Support Services

A digital preservation infrastructure will remain agile and robust by being able to respond to changes to information technologies, the evolution of user communities, and changing ways that users look for and use information. This will create a need in the infrastructure for institutions and individuals who are willing to take on the roles of support services. These include technology monitoring, staff training, and software development and maintenance.

Perhaps the most important long-term and long-range support service is advocacy. This includes designing and promoting laws and policies and advocating for funding.

Conclusion

The many-to-many model and the distribution of OAIS functions will provide the flexibility that libraries and archives need to apply their existing resources to the needs of the communities they serve. Collectively, the participation of many libraries will increase the amount of information preserved and multiply the digital user services for that information many times beyond what any individual library or archive can provide.

CHAPTER 23
Transition

We have suggested an open framework approach for developing a Digital Preservation Infrastructure (DPI) that would address existing preservation gaps. If, as Star and Ruhleder say, infrastructures have to emerge from practice, what actions can preservationists take today that will facilitate that emergence? No one can prescribe the path for the transition between the paper-era preservation infrastructure and a new Digital Preservation Infrastructure. Nevertheless, there are actions that preservationists can take today to move toward a successful DPI. In this chapter, we suggest concrete first steps.

Steps Toward an Ecology of Preservation

We suggest that the evolution of a new DPI will be a logical progression with each step building on the foundations laid by previous steps. The early development of a Preservation Plan, for example, will guide the first concrete actions. Those concrete actions will begin a steady increase in the amount of information being preserved. This will provide incentives for more libraries and archives to participate in the new infrastructure, which will result in more information being preserved. As more information is available in a stable, reliable, consistent system (the National Digital Archive of Government Information), memory institutions will be able to build collections and services for their users with assurance.

The steps we list here are suggestive rather than prescriptive, but we believe they are practical and would be effective. The early steps do not require any new laws or funding or obligations. Steps like these are essential. They are planning steps that express intention and actions that lay the groundwork for more substantive actions.

Step One: Acknowledge the Need for DPI

Preservationists, information producers, and information consumers will gather to acknowledge the need for a Digital Preservation Infrastructure. One or more key organizations (such as the Coalition for Networked Information, the Center for Research Libraries, the Depository Library Council, the Government Documents Round Table [GODORT], or the Preservation of Electronic Government Information [PEGI] project, etc.) will convene a meeting for this discussion. Participants will include government and non-government preservationists. This meeting will define the membership, structure and governance, and agenda of a Preservation Planning Committee that will provide guidance for digital preservationists.

Neither the convening of this initial meeting nor the creation of a Preservation Planning Committee requires any new laws or official approval. These are simply people with common interests convening to discuss problems and solutions. Such actions have precedent. For example, in 1996 the Research Libraries Group (RLG) and the Commission on Preservation and Access convened a task force that suggested the need for a network of trusted and certified digital archives (Waters & Garrett, 1996). And in 2002 RGL and the Online Computer Library Center established a working group to define attributes of trusted digital repositories, which evolved into the ISO standard 16363, *Audit and certification of trustworthy digital repositories* (ISO, 2012a) (see Appendix A).

Step one would result in the library and archive community agreeing on the broad outlines of a DPI and the creation of a DPI Preservation Planning Committee.

Step Two: Develop a Preservation Plan

The first task of the DPI Preservation Planning Committee will be to define the goals and strategies within which libraries and archives can act. The goals and strategies will be designed to work within the context of existing legislation, responsibilities, and authority. The goals and initial strategies will provide the guidance for government and non-government preservation institutions to choose independent actions that result in collaborative preservation. The Plan will guide preliminary actions and provide a framework for future actions built on that foundation.

In Chapter 19, we outlined elements of a goal by defining Public Information and preservation, and by enumerating the reasons for preservation.

Here we suggest three strategies for working toward a common goal of preserving the federal government's digital Public Information. (Note that strategies are distinct from objectives and tactics and tools. See Chapter 19.)

1. Define a foundation of permanent, distributed storage and persistent, uninterruptible availability at sufficient scale to deal with the quantity of born-digital federal government Public Information being produced. (This will provide adequate permanent storage for a comprehensive National Digital Archive of Government Information and a basic level of free public availability of that information. This will be the foundation on which the other elements of a DPI can evolve.)

2. Establish standards, practices, and procedures for distributed selection, acquisition, and ingest of digital government Public Information. (This will increase the capacity for ingest to match the necessary scope and coverage for building a comprehensive archive of government Public Information.)

3. Establish an implementation committee to facilitate, guide, and coordinate actions of participants.

Step Three: Begin Implementation

The Planning Committee will create an Implementation Committee. Its task will be to develop objectives and tactics for implementing the strategies and to specify standards for use in doing so. With those in place, individual memory institutions will be able to align their existing and planned tactics and tools with the near-term objectives and long-term strategies of the DPI. At this point, a few government and non-government institutions will offer to be Primary Archival Repositories, and decisions will be made to adopt a multiple-resolution Persistent Identifier (PID) technology.

Existing preservation projects at government agencies such as NARA, GPO, and the Library of Congress will probably need little modification in ingesting, storing, and making information available to the public as Primary Archival Repositories. The actions at this stage will mostly be adapting existing tactics and tools and policies to specific DPI objectives as specified by the Implementation Committee. Similarly, an existing project such as LOCKSS-USDOCS could quickly align its actions and procedures with the new Preservation Plan to become the first Replication Repositories.

The first major new procedure will be the adoption of a PID technology and implementing its management. Using multiple-resolution PIDs (URLs that point to more than one copy of the same information object) will be a major new task. It will ensure that at least two copies of each preserved digital object are stored in at least two separate repositories and that all stored copies of any given digital object can be retrieved via a single PID.

The early stages of implementation will provide an opportunity to experiment with procedures and responsibilities that can eventually scale up to a comprehensive digital archive that spans multiple Primary Archival Repositories. It will, therefore, be implemented incrementally in order to develop and refine those procedures and responsibilities that can best scale up.

A flexible design of standards and requirements will enable existing programs and projects to align with a common goal.

Step Four: Laws

A new infrastructure will not be designed by laws, but laws can definitely facilitate or impede a new infrastructure. As described in Chapter 5, laws can interfere with preservation if they are out of date or too specific or badly crafted. Because laws change slowly and technology changes rapidly, the best laws are not technologically specific.

Laws are also difficult to enforce. It is often all too easy for agencies to ignore or work around laws and regulations that they find inconvenient or expensive. (See, for example, the story of the paper-era preservation gap in Chapter 5 and the story of the *Government Information Locator Service* in Appendix E.) It is also easy for Congress to fail to fund implementation adequately.

It is also important to recognize that bad laws are worse than no laws. Attempts to modernize existing laws can cement new technologies to old infrastructures (see Chapter 5). And the legislative process of writing new laws or updating old ones can attract new attempts to privatize or commercialize government information (see Chapter 1).

Perhaps the best laws for information preservation are those that follow existing good practices rather than those that try to create good practices. Good laws can instantiate and formalize existing practices and accepted norms rather than initiate them. For example, libraries existed before laws relied on their existence to create the Federal Depository Library Program. Just as libraries did not need a law to tell them to catalog their books or what standards to use to do so, so digital libraries and archives do not need laws to enable them to follow a goal-driven preservation plan and use shared standards, practices, and principles. Laws, regulations, and official policies can, however, simply require agencies to preserve their published Public Information. The policies of the National Science Foundation (NSF) and National Institutes of Health (NIH) and other agencies that require making the published results and data gathered with federally funded research publicly available are a good example of this (Pasek, 2017). Such requirements create a feedback loop as well. As more information needs preservation, more libraries and archives participate in the preservation infrastructure (Fearon et al., 2013; Tenopir et al., 2015; Boté-Vericad & Healy, 2022; Safdar et al., 2022).

The time for new laws is when a digital preservation infrastructure has begun to emerge. After the steps described above have established a goal and strategies and after libraries and archives have started implementing specific tactics to achieve objectives within a common framework, a new law can provide agencies the additional legal incentive they need to participate in (and benefit from!) that infrastructure.

With a nascent DPI in place, libraries and archives and other interested communities will design and lobby for laws and policies that will specifically require agencies to preserve their Public Information. The precise wording of such laws and regulations would parallel the goal of the DPI itself, thus encouraging agencies to seek out DPI solutions.

Conclusion

A preservation plan based on an open framework of transformative ways of preserving information can enable participation of many libraries and archives toward a common goal. As more libraries and archives participate, an infrastructure will evolve, making it easier and more productive for more libraries and archives to participate. The results of this will be more government information preserved more securely with more ways of discovering and using that information by more communities.

CHAPTER 24
Conclusions

When we began writing about the issues in this book, the risks of losing born-digital government information were mostly theoretical (Free Government Information, 2025). Any frequent user could easily point to born-digital information being moved or altered or lost, but often—even if NARA could not find a copy—one could find copies stored somewhere on some non-government site—if one was willing to spend enough time hunting (Jacobs & Jacobs, 2021). There were cases of temporary interruptions of service (Jacobs, 2009b) and temporary withdrawals of information (Jacobs, 2013), but an optimistic preservationist could reasonably think that the preservation of government information was, if not perfect, at least "good enough."

But now there are two things that make such optimism demonstrably unrealistic. First, there is the data analysis we present in Part Two of this book, which documents large gaps in the preservation of government information. Although the documentable gaps are particularly severe for the executive branch (the largest government publisher), the lack of a digital preservation infrastructure puts information from all three branches at risk of loss. These gaps demonstrate that relying on the government to preserve its own information is just not working. The good news is that the gaps documented here are, for the most part, unintentional gaps caused by an inadequate infrastructure (Jacobs & Jacobs, 2025).

The second thing that makes this optimism unrealistic is the active alteration, destruction, and withdrawal of information carried out by the Trump administrations in 2017 and 2025 (Gehrke et al., 2021; Shendruk & Rampell, 2025). These overt actions to skew or delete the public record demonstrate the inherent vulnerability of relying on the government for preserving government information. This is very bad news. Actions like these change the problem from the government failing to provide adequate preservation, to the government being an active enemy of preservation.

Although the second Trump administration's actions have been unprecedented in scale, it is not the only administration that has attempted to control, restrict, or suppress government information. Similar problems of access have a long history stretching back to the paper era. In the ALA publication *Less Access to Less Information by and about the US Government*, Anne Heanue documented such problems from 1981 to 1998 (Heanue, 1981). And patterns of control and suppression have been particularly evident since 2001 (Jacobs, 2002; Peterson & Jacobs, 2005; Jacobs, 2016).

The unavoidable conclusion from all this is that preservationists cannot rely solely on the government to preserve its own information. Its actions have been inadequate at best and, at worst, actively hostile to preservation.

This is big news because, for years, libraries have relied on the government to preserve its own information. Preservationists have cited with confidence the principle that the government has an obligation to do so (Chapter 1). But the weaknesses in the laws that govern preservation have led to unenforceable policies (Chapter 5) and big gaps in preservation (Part Two). Even when, in the paper era, preservation relied on hundreds of Federal Depository Libraries outside of the government, that system also had gaps that are attributable to a system that was defined—and limited—by laws, and administered by a government agency (GPO). That paper-era infrastructure of outside-the-government preservation is gone now, replaced by GPO's digital preservation repository, but the new system suffers from the same kinds of limitations. The bulk of content in GPO's GOVINFO are traditional Congressional publications and an incomplete duplicate of court records that could be freely available on the web if the courts weren't making so much money by keeping the complete record behind the PACER paywall (Chapter 7).

Another unavoidable conclusion is that, in the digital age, even the dedicated preservationists within the government (and there are many and we applaud them and their excellent accomplishments!) are hampered from preserving government information comprehensively by the government's own outdated laws, restrictive rules, and inadequate funding.

In the early days of the web, there was some hope that agencies producing government information might keep their own published information online, but that hasn't happened. We know now that agencies alter, replace, and remove information all the time. Link rot and content drift are endemic to the web, and the government web is no different (Chapter 6).

And now we have demonstrable proof that, when executive branch that doesn't care about preservation—or when it wishes to actively alter and destroy information, it can quickly erase and alter information at scale. As one study concluded after the 2017 crisis, there is an "absence of a legal framework to protect, preserve, and ensure access to publicly-funded digital information, web content may be altered, removed, or otherwise hidden without substantive repercussions" (Nost et al., 2021). Each time this happened, it prompted valiant attempts by volunteers to try to save information before it was deleted. Preservationists must now acknowledge that the only way to guarantee that government information will be adequately preserved is for non-government preservationists to acquire and preserve government information as it is published, not as it is being destroyed.

As a practical matter this means that, rather than assuming that government will take all responsibility for preservation, librarians and archivists outside of government have to take primary responsibility for the preservation of government information. Perhaps, one day, Congress will put the principle of the government's responsibility for preservation into practical laws. But, as we write this, that day seems, at best, very far off and, at worst, impossible.

The preservation community should not ignore government agencies, however. Preservationists should insist that government agencies take on the roles that they can uniquely fulfill, such as publishing digital information that is Findable, Accessible, Interoperable, and Reusable (FAIR) (PEGI Project, 2024). The government is also in the best position to create metadata and ensure that DPI participants can easily and comprehensively identify, describe, acquire, and preserve information released to the public (Chapter 22). These are technically and financially easy targets for change.

Reversing the reliance on the government will require big changes in library policies. If the first quarter-century or so of the digital age can be characterized by libraries saying that government information was not their responsibility, the next quarter-century needs to be libraries acknowledging that it is their responsibility. What are the incentives for them to do that?

The Loss of Old Incentives

The key question that library and archive managers face is how to allocate their limited resources. Whether this has to do with collections, or staffing, or physical facilities, or services, it is the key management challenge of libraries and archives. If, as we suggest, the future of the preservation of government information will depend on the participation of many libraries and archives, will managers approve of expending resources on preserving government information? The history of library support for the depository program provides mixed evidence.

In the early days of the digital age, some managers questioned the wisdom of participating in the FDLP unless someone provided them with "incentives." In 1993, for example, as GPO was building its first digital repository, a report examining alternative structures for the depository library program repeatedly asked the question, "What are the incentives?" Regional depositories in particular complained that the "obligations" that GPO "imposed" on them were "too burdensome and inflexible," and they argued that their job was not to serve as "archives" of government information (Depository Library Council, 1993).

By 2003, the Depository Library Council (DLC) identified "benefits" that GPO could provide "to encourage depositories to remain in the Federal Depository Library Program" (Depository Library Council, 2004). Some of these benefits were supposed to enhance public access, but others were aimed directly at providing depositories with a "better return on their investment." These included funding and subsidies and policies that would make it easier for depositories to discard government information. Such approaches are common in the transition from an old infrastructure to a new one; they attempt to preserve a tactic (in this case, the FDLP), rather than fulfill a goal (in this case, preserving government information).

One of the biggest incentives for libraries to join the FDLP in the paper era was that they got "free" books, which increased their volume count (Housewright & Schonfeld, 2011). This was a benefit to libraries that used collection size as a key indicator of the quality of the library. A large volume count raised a library's profile in library associations like the Association of Research Libraries (ARL), for example. But by 2007 libraries were de-emphasizing collection size as a measure of quality. Significantly, ARL changed the way it measured its member statistics by de-emphasizing collection size (Payne, 2007). By 2008, when GPO surveyed the directors of Regional Depository libraries, it found 20 percent of them were considering dropping their regional depository status. Ithaka S+R's 2009 study of the depository program concluded that there was "little incentive for many of the largest research-oriented libraries to remain in the Program" (Schonfeld & Housewright, 2009a).

This emphasis on incentives was, we believe, a reflection of the failure of the paper-era infrastructure. After the *GPO Access Act* was passed, GPO essentially stopped depositing government information with depository libraries (Chapter 5). This meant that, for new publications, libraries' role of providing both short-term and long-term access was gone.

With no new infrastructure in place, it was understandable for depository libraries to question their support for the old, paper-era infrastructure, including the FDLP itself. This led to a spate of papers and reports about options for the FDLP, some of which questioned its very existence. (For example, "The reinvention of the FDLP" [Arrigo, 2004], "The Federal Depository Library Program in 2023" [Hernon & Saunders, 2009], and "Modeling a Sustainable Future" [Housewright & Schonfeld, 2011].) What did not emerge from all this speculation and discussion was a vision of new infrastructure. What was needed was a new digital preservation infrastructure that provided libraries opportunities for meeting the needs of their communities through participation in the infrastructure.

New Incentives

We believe that the open framework we have outlined here provides just such opportunities. It provides a positive incentive for libraries to participate that contrasts sharply with those calls for subsidies and the complaints about obligations.

In a 2003 report on the incentives to preserve digital materials, Brian Lavoie of OCLC acknowledged that some institutions might require a "pecuniary" incentive. But he noted that this was not the only reason for participating in digital preservation. Some institutions, he said, will be motivated to preserve information by an "altruistic desire to ensure the availability of digital materials for research and education" or a "perceived responsibility to perpetuate the cultural record" (Lavoie, 2003).

We suggest that the incentive that libraries need to participate need not be either pecuniary or altruistic or even an "obligation" imposed by an inflexible bureaucracy. A successful DPI will replace obligations with opportunities for libraries to provide better and more secure services to their users. A successful DPI will also replace the call for subsidies with efficiencies and cost savings. We suggest that libraries will participate in a DPI because it will become evident that doing so is the best and most efficient way to fulfill an essential part of their existing missions and their obligations to their users. The "incentive" for participation will be positive, demonstrable results for the communities that libraries serve.

For library managers to reach that conclusion, they will have to acknowledge that government information has an inherent and irreplaceable value, that their constituencies need that information, and that the government cannot ensure the preservation and availability of this information by itself. In short, they will have to acknowledge the need for a Digital Preservation Infrastructure and embrace the role of libraries and archives in that infrastructure. That will be both a practical and a philosophical justification for participation.

But a DPI with a many-to-many model of collections and services and a distributed implementation of OAIS will also provide an economic benefit to participants. Every dollar invested will go further as part of a collaborative endeavor than it would go as a project that benefited only a single institution. Every dollar a library invests will contribute to the creation and preservation of a more comprehensive National Digital Archive of Government Information that will be accessible and usable to its constituency. And every dollar invested in the DPI will contribute to a network of universally accessible defined collections and specialized digital services. A library that can only afford to create a single service or contribute to one small part of one service will nevertheless be part of the collectively built infrastructure of preserved, accessible, usable government information. Their participation will be part of a new, transformed culture of preservation and access.

The Future is Now

Even without a DPI in place, there is good evidence that librarians and archivists want to have a role in preserving government information.

One obvious indicator is the volunteer efforts to rescue government information in 2017 and 2025. Groups of volunteers such as the Public

Environmental Data Partners (Public Environmental Data Partners, 2025) and Data Rescue Project (Data Rescue Project, 2025b) have donated massive amounts of expertise, work, and technological infrastructure to preserve government information. The End of Term Archive has successfully preserved over a petabyte (1,000 terabytes) of data since 2008. Besides EOT's official partners, hundreds of individual volunteers submit seeds for preservation. For the 2024 EOT crawl, 432 individuals submitted a total of more than 12,000 seed nominations.

Those examples are from crisis mode, but they build on an existing foundation of volunteers who, even without an underlying DPI, have created collections and services and forged collaborative partnerships to ensure long-term access to government information. A review of surveys of FDLP libraries showed that hundreds of FDLs have been downloading and storing government information on library servers at least since 2005 (Jacobs & Jacobs, 2017).

These digital-age actions have their roots in the paper era. These include projects such as the Documents Expediting Project at the Library of Congress (Shaw, 1966) and the long battles to get CRS reports (Young, 2006) and SEC reports (Love, 1993) into public repositories.

In the digital age, there are many collaborative projects inside and outside the government. These include GPO's collaborations with NARA (GPO, 2023c) and GAO (GPO, 2025); GPO's "partnerships" as a substitute for Regional Depositories (FDLP, 2025a); a cataloging collaboration between the University of Montana and GPO (Keenan et al., 2013); the "Department of State Foreign Affairs Network" partnership between the State Department and the University of Illinois at Chicago (US Department of State, 1995); the Technical Report Archive & Image Library (TRAIL) (Center for Research Libraries, 2025); the Internet Archive's "Democracy Library" (Kahle, 2022); and the non-profit USAFacts, which makes government data easy to access and understand (USAFacts, 2025). Cornell alone hosts several projects including the feature-enriched collection of primary law (Legal Information Institute, 2025); the USDA Economics, Statistics, and Market Information System, which, in partnership with Department of Agriculture, provides access to historical publications (Mann Library, 2025); and the collaboration with the US Department of Labor for archiving labor reports (Davis, 1996).

Preserving large datasets presents big challenges, which have been addressed by big projects. These include the University of California's Datamirror project (University of California Curation Center, 2018), ICPSR's collaboration for hosting criminal justice, aging, census, and other US data, and its DataLumos crowd-sourced repository for valuable government data (ICPSR, 2025); and Harvard's mirroring of data and metadata held and described in data.gov (Harvard Library Innovation Lab Team, 2025).

On a smaller scale are the dozens of individually maintained "LibGuides" that focus on government information (FDLP, 2025b). These are carefully maintained collections of links to government information by agency, subject, and discipline. Imagine if the detailed work that goes into producing these guides could result in content being ingested into a National Digital Archive of Government Information with every item having a PID and being professionally preserved.

All these examples are notable because they have emerged in spite of the fact that there is no underlying digital preservation infrastructure to support their work or to which they can contribute. This shows that there is a willingness—even an eagerness—on the part of government information librarians to do more to make government information discoverable and usable for the long term. These librarians need an infrastructure that will support and amplify their work beyond their individual projects and individual institutions. The task of those who care about preserving government information is to build on that untapped eagerness.

For those who value preserving government information, it should be clear what a successful Digital Preservation infrastructure must do.

- It must support sharing and interoperability of collections and services natively.

- It must support libraries and archives that share content and manage complementary services.
- It must do more than make static digital objects available. It must support new kinds of discovery, identification, acquisition, and use of information itself by new kinds of user communities.
- It must enable libraries and archives to select and acquire Public Information precisely and comprehensively so that they can provide their own information services for discovery acquisition and use of that selected information to their Designated Communities.
- It must support libraries and archives that provide services for collections that they can select and control.
- It must support PIDs as essential, not an optional convenience.
- It must guarantee that information will be:
  - Not just preserved but discoverable. [OAIS: 2.2.2]
  - Not just discoverable but deliverable. [2.3.3]
  - Not just deliverable as bits but readable. [2.2.1]
  - Not just readable but understandable. [2.2.1]
  - Not just understandable but usable. [4.1.1.5]

The task we must adopt is to create a digital preservation infrastructure for the comprehensive preservation of government information—starting today.

APPENDICES

Appendix A:
Certification and Sustainability

Costs and Benefits of Using OAIS and TDR

This book uses the OAIS standard, ISO 14721 (ISO, 2012b), as a key to organizing the preservation of government information, but many people assume that OAIS is difficult and expensive. If this assumption is accepted, it can be used to justify avoiding the use of OAIS for planning as a "cost-saving" measure. We believe that the assumption is wrong and the cost-savings are illusory.

Most of the concerns about the cost of conforming to OAIS are not directed at OAIS itself but to the cost of becoming certified as complying with OAIS (Alliance for Permanent Access to the Records of Science Network, 2012; Corrado, 2019; Lippincott, 2018).

The standard for certifying that a repository conforms to OAIS principles is ISO 16363, *Audit and Certification of Trustworthy Digital Repositories*, which is sometimes referred to simply as "TDR." In our experience, the cost of certification is inversely proportional to the amount of OAIS planning the repository has used. If a repository has not used OAIS in planning and developing a repository, it may well find that the costs it incurs during the certification process may be high. It may, for example, have to map its existing procedures and policies to OAIS terminology and functions after the fact. That process might be confusing to the OAIS novice and, so, be expensive. It may also reveal gaps in preservation planning, which may require more work and more costs. But if a repository has used OAIS as a basis for developing a preservation plan, the certification process will not be burdened by such additional costs.

An understanding of the costs of using OAIS for planning should, therefore, be understood in two parts. The first is the cost of using OAIS for preservation planning from the start. Planning to include all OAIS functions and using OAIS terminology for planning should make preservation itself more reliable and, therefore, less costly than not doing so. The second cost is optional: seeking TDR certification. If a repository chooses to seek certification, the cost of doing so will be much less if the repository has used OAIS for planning from the start. But not every repository will choose to get certification and so will incur no additional costs.

Thus, the assumption that conforming to OAIS makes preservation more expensive is wrong for two reasons. First, a repository does not have to get TDR certification, but those that do will find the process easier and less expensive if they have used OAIS for preservation planning. Second, failing to take OAIS functions into account during planning can create more expenses later when preservation gaps are discovered.

This does not mean that ISO 16363 is irrelevant to preservation planning. On the contrary, it can be helpful to use its metrics during planning to help understand, clarify, implement, and document OAIS functions.

History of TDR

As preservationists consider the feasibility of moving to a new Digital Preservation Infrastructure with a distributed implementation of OAIS, it is useful to understand TDR's strengths and the relationship between OAIS and TDR.

OAIS, which was originally written in the late 1990s, did not include specific metrics for evaluating how a repository could be trusted to provide the functions and meet the responsibilities that it defined. It did, however, suggest the need for such standards. It specifically recommended that additional standards should be developed for archival practices and for the accreditation of archives (Consultative Committee for Space Data Systems [CCSDS], 1999).

Such a standard would give a repository a way of independently certifying to the world that its practices are reliable and that it is OAIS compliant. It would also provide a standardized way for an archive to specify for whom it is preserving information and why and how it is doing so. Finally, it would assure its users that it is preserving their data securely.

ISO 16363 (TDR) was approved in 2012 and reviewed in 2023 (ISO, 2012a). It defines a recommended practice for assessing the trustworthiness of digital repositories. While it was designed as a tool for certifying archives, it can also be used as a planning tool to guide an archive in the creation of a new repository or in extending the capabilities of an existing repository (Baucom, 2019). It can also help an archive evaluate its existing procedures, identifying those that need to be modified when its environment changes.

The creation of TDR has a long history that begins at least as early as 1996. In that year, a task force of the Research Libraries Group (RLG) and the Commission on Preservation and Access (CPA) issued its final report, *Preserving Digital Information*, which suggested the need for a network of trusted and certified digital archives (Waters & Garrett, 1996; Commission on Preservation and Access, 1996; Research Libraries Group, 1997).

RLG and OCLC (then known as the Online Computer Library Center) followed up on the RLG/CPA report by setting up a working group to establish attributes of trusted digital repositories based on OAIS. In 2002, RLG and OCLC issued the working group's report, *Trusted Digital Repositories: Attributes and Responsibilities* (Research Libraries Group, 2002). That report defined a trusted digital repository as "one whose mission is to provide reliable, long-term access to managed digital resources to its designated community, now and in the future." That definition was somewhat narrower than that of the OAIS standard, which was ready for approval by ISO in 2003 (ISO, 2003). RGL/OCLC wanted to ensure the sustainability of the archive itself for the long term so that it could preserve the information for which it took responsibility. In contrast, OAIS envisioned the possibility of information being preserved even in the face of institutions failing. OAIS focuses on the long-term accessibility of the information rather than on the survivability of individual institutions. It requires institutions to have succession plans (OAIS 3.3.6) but does not require individual institutions to be "sustainable." OAIS's broader definition implicitly accommodates the possibility of temporary staging repositories (Steinhart et al., 2009; Khan et al., 2011) and other ways to distribute implementation of OAIS (OCLC Online Computer Library Center Incorporated, 1999).

The RLG/OCLC report suggested that trusted repositories must be required to demonstrate administrative responsibility, organizational viability, and financial sustainability. It specifically called for a trusted repository to be able to prove its financial sustainability over time and to have a sustainable business plan. It enumerated "attributes" and "responsibilities" and set out recommendations for further study, but it did not provide specific criteria for assessing and measuring trustworthiness.

A year after the RLG/OCLC report was issued, RLG and the National Archives and Records Administration (NARA) created a task force to develop specific criteria for trusted digital repositories. Drawing on the suggestions of the RLG/OCLC report and the work of the Digital Curation Centre, the Australian Partnership for Sustainable Repositories project, the National Library of Australia, Germany's nestor project, and others, the RLG-NARA Task Force on Digital Repository Certification specifically recognized that reliable, long-term access to digital resources was about more than just a digital preservation system. The task force determined that to guarantee trustworthiness, "one must look at the entire system in which the digital information is managed, including the organization running the repository: its governance; organizational structure and staffing; policies and procedures; financial fitness and sustainability." The result of the task force's work was the first set of criteria for trustworthy digital repositories, the *Trustworthy Repositories Audit & Certification (TRAC): Criteria and Checklist Version 1.0* (RLG-NARA Task Force on Digital Repository Certification, 2007).

TRAC enumerates 84 specific criteria in three broad areas: organizational infrastructure, digital object management, and technology. The criteria are specific without being prescriptive and include examples of the kind of evidence that might be used to demonstrate compliance with the criteria.

Where OAIS provides a framework, TRAC provides an actual checklist of criteria. For example, the OAIS Information Model describes Preservation Description Information (PDI), which includes integrity checks on digital objects in the form of "fixity information" (e.g. checksums). TRAC translates this general function into five specific, measurable criteria that mention fixity and integrity.

Following the publication of TRAC, the Consultative Committee for Space Data Systems (CCSDS), which had created OAIS, began work on a more formal version of TRAC. This work resulted in the publication of ISO 16363:2012, *Audit and certification of trustworthy digital repositories* (ISO, 2012a). Where TRAC has 84 criteria, TDR has 109 criteria that it refers to as "metrics." Metrics were added, removed, and changed from TRAC. The bulk of the changes were to the section on digital object management. For example, where TRAC mentions "fixity" in 5 criteria, TDR mentions it in 10 metrics. TDR ISO-16363 is, however, a refinement of TRAC, not a wholesale change.

Appendix B:
Depository Policy

In this appendix, we examine the GPO policy for distributing government information to Federal Depository Libraries and that policy's assumptions and the effects of the policy on the development of a Digital Preservation Infrastructure. We examine this policy as a kind of case study of how clinging to out-of-date assumptions and an out-of-date infrastructure can hinder the development of a successful infrastructure.

As agencies started making their information available directly to the public on the web, the paper-era infrastructure that preserved government information by providing both short-term and long-term access in libraries broke. A new infrastructure was needed if government information was to be successfully preserved. Deposit of government information with depository libraries was a key component of the old infrastructure. Changes to that policy had the potential to either enable or hinder the development of a new digital preservation infrastructure.

History of the Policy

1980s

GPO's policies that govern the distribution of information to Federal Depository Libraries (FDLs) have gone by various names and numbering schemes and have been periodically revised over the years. The first significant change to the depository distribution policy came in the early 1980s when GPO's Superintendent of Documents (SOD) had to revise it to accommodate microfiche. (The Superintendent of Documents is the office in GPO that administers the depository library program.) The new policy was then known as "Format of Publications Distributed to Depository Libraries" and was designated as "SOD 13" (GPO

Superintendent Of Documents, 1983). Although it fit in easily to the paper-era infrastructure of depositing physical objects, it took a long time to implement. GPO had asked for permission to publish microfiche more than 10 years earlier. Distribution had started in 1977. By 1981, microforms were the "future of distribution," according to GPO (Sleeman, 2009).

The adoption of microfiche as a distribution format seems, in retrospect, like a small change, but even small changes can be difficult to implement in big bureaucracies. There were extensive discussions about appropriateness of adding microfiche to the depository program. There were discussions of how it would affect the costs to GPO (microfiche publications would be less expensive to produce and distribute) and to libraries (libraries would save space but would have to acquire cabinets and fiche-readers). In deciding which publications to convert to microfiche, GPO also considered some user needs, leading to a decision to keep nontechnical publications on subjects like gardening and childcare and publications intended for the general public, high school students, homeowners, and the elderly in paper format. The use of microfiche brought more publications into the program because of the cost savings, but it also made many heavily used publications less convenient to use when they were distributed on fiche to save printing cost (Kessler, 1996). As for preservation, there was a controversy over whether or not to require the use of archival quality silver halide fiche or less expensive diazo fiche. The decision was that nothing justified the additional expenditure of providing archival quality fiche to depositories (Schwarzkopf, 1978).

The 1983 policy change was a practical adaptation of the paper-era infrastructure. It was based on lowering costs, saving space, and making more documents available—but at some sacrifice of convenience for users. The preservability of microfiche in FDLs was considered, but archival-quality fiche were rejected because of their expense.

1995

The next big change that FDLP faced was how to accommodate digital publications. In the paper era, FDLs were responsible for access to and preservation of published government information, but in the digital age, GPO would take those responsibilities away from them, arrogate them to itself, and provide no new role for libraries. The shift to digital publishing certainly required changes to how government information would be preserved, but GPO used paper-era assumptions about access, preservation, and the nature of information itself to develop its response.

GPO introduced this change in a draft of a plan to supersede SOD 13 in its 1995 *Transition Plan* (GPO, 1995b). It described the *Plan* and the policy as its "tactical" response to the government's shift to digitally published Public Information. This tactical plan was to be superseded by a longer-term "strategic" plan within a year, but the long-term strategic plan adopted the same approach as the short-term tactical plan (GPO Superintendent of Documents, 1995; GPO, 1996a). Through various iterations of the policy and changes in terminology over the years, this policy change has remained constant and thus has perpetuated the redefined preservation roles of GPO and FDLs to the present day (GPO Superintendent Of Documents, 2001; 2005; 2006; 2008; 2019b). The most recent version of the policy has even dropped "publications," "distribution," and "depository libraries" from the name of the policy.

The policy that took "deposit" and "libraries" out of the depository library program was a response to the fact that the paper-era infrastructure wasn't designed to accommodate digital information. The shift to digital publishing made it clear that a change in policy was needed.

Although digital publishing brought with it many significant changes to the inherent characteristics of information (Chapter 16), it was two other issues that dominated the policy discussion in the late 1990s. The first was that agencies were increasingly publishing information directly on the web rather than in paper. When GPO could get copies of such information, there was no infrastructure in place to distribute it to depository libraries. As we described in Chapter 14, few depository libraries had adequate digital infrastructure to select, accept, store, or provide digital services for online information in the mid-1990s. In the face of that, the question the problem prompted was how to ensure long-term access to this information.

The second problem was that some agencies were distributing digital information on floppy discs and CD-ROMs. Since few FDLs were adequately equipped to house and provide public services for such disks, the problem prompted the question of how to deal with information on those disks.

Although the digital shift raised other questions, particularly about preservation, GPO had resources to respond to these two questions. It did so and designed the policy around that response.

First, as required by the 1993 *GPO Access Act*, GPO had a digital storage facility with online access, and it could use that to "disseminate" (if not "distribute" or "deposit") digital files in its possession. Second, GPO had the paper-era infrastructure for shipping physical objects to FDLs. It could mail disks as easily as it mailed books.

By dividing all digital information into two simple categories, GPO could address both of those immediate problems of making digital information available. GPO called information that it could physically ship (such as CD-ROMs and diskettes) "tangible electronic products." Everything else was "intangible electronic data" that would be "available...via remote electronic access." This response was a practical approach that enabled GPO to take action quickly.

But this approach failed to address all those other questions that the digital shift raised. It failed to take advantage of the unique characteristics of digital information. It neglected long-term planning for preservation and access in favor of a short-term approach that simply used the existing infrastructure of FDLs handling physical objects and GPO storing static digital objects. In short, it failed to address the long-term problems, challenges, and potential advantages of the digital shift.

Instead of working toward a new infrastructure for the preservation of digital information that included enhanced access and usability, the policy used paper-era assumptions and the infrastructure that was already in place.

Below we examine those assumptions and the effects of the policy.

"tangible electronic products"

The policy for "tangible electronic products" accepted the paper-era assumptions about physical objects and local access and used the paper-era infrastructure for shipping those physical objects. In this way the policy dealt with the problem much the way the 1983 policy had dealt with microfiche: it just added disks as an acceptable media for transmitting information to depository libraries. It was based on the paper-era assumption that libraries could adequately provide short-term and long-term access to the information on those disks just as it provided short term and long term access to printed publications and microfiche. That assumption proved false.

Just as the 1983 microfiche policy change had required some slight modification of the installed infrastructure (libraries had to acquire fiche cabinets and fiche readers), so the "tangible information products" policy required modification to the installed infrastructure. Many FDLs did not have the hardware and software for dealing with those deposited disks. The information on the discs was often stored in proprietary formats that required specific (sometimes commercial) software to use. The software required specific operating systems. Even the physical discs themselves required specific hardware and drivers to be readable. In order to adapt the old access model to the digital age, FDLs would need the infrastructure (hardware and software and procedures and training) for providing access to the digital information GPO was depositing with them. GPO, therefore, issued technical requirements for depository libraries that included requirements for operating system, hardware, and commercial software (GPO Library Programs Service, 1995). Eventually hundreds of floppies and thousands of CDs were distributed to depositories (Gano & Linden, 2007; Woods, 2010).

The policy specified that only digital information written to mass-produced disks would be deposited with depository libraries. In 1995, few FDLP librarians or GPO staff recognized the long-term access problems of basing a policy on distribution media rather than on preservability of the information on those disks, but it soon became difficult to ignore those problems. Even before the policy was written, GODORT had reported that the usable lifespan of the disks themselves was unknown but assumed to be short (GODORT Federal Documents Task Force, 1992). In 1996, NASA issued a report on its experiences and requirements for long-term data storage in which it said its "philosophy has always been to preserve the data and not the medium onto which it is written" (Holdsworth, 1996). That same year, GODORT reported that NARA, which accepted CD-ROMs as a "transfer medium," was copying the data on those discs to magnetic tape cartridges for permanent storage because of the lack of standards for the archival quality of CD-ROMs (GODORT Ad Hoc Committee On The Internet, 1996). In a 1999 report commissioned by GPO, NCLIS noted specifically that there was "a lack of standardization for producing Government information products on CD-ROM" (US National Commission on Libraries and Information Science, 1999). In September of 2000, there were brief discussion on the government documents mailing list about problems with the Census "test disk 1" and its "data format issues" (govdoc-l, 2000).

These isolated indicators were part of a growing understanding of digital preservation issues in general. The first version of OAIS was released in 1999, providing the terminology needed to articulate the issues (Consultative Committee for Space Data Systems [CCSDS], 1999).

By 2000, distributing information on mailable disks had become less popular. Disks distributed to FDLs in 2000 accounted for only about a tenth of a percent of information distributed (Baldwin, 2000).

In 2004, John Hernandez and Tom Byrnes, in a presentation to Depository Library Council, spelled out the "CD legacy problem" of obsolescence and lack of backward compatibility and outlined a project to quantify the problem and address it (Hernandez & Byrnes, 2004). One of the possible solutions that they listed was to use the recently announced GPO "collection of last resort" (Baldwin, 2003; GPO, 2004a) to create a "[s]torehouse of content available for possible re-formatting." This would be an ironic reversal of the 1995 policy of assigning FDLs responsibility for long-term access to "tangible" products because it would reallocate that responsibility to GPO.

There followed small projects at universities to try to save the information on floppy discs and larger projects to address preserving the information contained on thousands of CD-ROMs (Gano & Linden, 2007; Woods & Brown, 2009; Woods, 2012). GPO launched a pilot project to test migration processes for CD-ROMs in 2006, an "assessment project" in 2009 to identify discs at risk for content loss (Davis, 2009), and in 2011 a project to look for "equivalent versions" of CDs on government web sites (FDLP, 2011). More recently, libraries have explored the idea of emulation as a service with the EaaSI project (Software Preservation Network, 2024).

It is not clear how much of this information was lost and how much was saved. What is clear is that the "tangible" part of the 1995 policy had largely failed as a preservation (or "long-term access") strategy for the information on those disks. That failure also demonstrated that preserving digital information was going to require replacing the paper-era infrastructure with an explicit long-term digital preservation strategy.

"intangible electronic data"

To its credit, the 1995 policy did take one of the new characteristics of digital information into account. It recognized that, where paper publications had to be physically distributed to libraries to make them easily accessible throughout the country, digital files did not have to be stored locally to be accessible locally. To accommodate this characteristic, the new policy dropped the paper-era model of access-through-deposit and replaced it with online access. To accomplish this, the *Plan* explicitly shifted the responsibility for ensuring long-term access to electronic information "from the depository libraries to the Superintendent of Documents." This shift was to be permanent and complete since it was designed with the assumption that "nearly all of the information provided through the Federal Depository Library Program (FDLP) will be electronic by the end of fiscal year 1998."

GPO was right to acknowledge the value of "remote electronic access," but its policy to respond to this valuable characteristic of digital information made two mistakes. First, it treated digital information the same way it treated printed publications. In other words, it adopted the paper-era assumption that making static objects available was sufficient for digital information. As we described in Chapter 17, this is a bad assumption that ignores the other inherent qualities of digital information. Second, by ceasing deposit, it took away the paper-era preservation and access roles of depository libraries without replacing those roles with a digital-age role. This deprecated the value of Federal Depository Libraries as partners in preservation and access at the very moment GPO needed strong, active, committed partners. It didn't take long before FDLs started questioning why they should stay in the program at all (Kram, 1998). (See also the section on "incentives" in Chapter 23.)

The policy also used paper-era assumptions about preservation and those assumptions allowed it to avoid developing an adequate digital preservation infrastructure. The policy explicitly relied, for example, on the paper-era division of responsibilities between NARA, which the policy asserted would be responsible for "permanent preservation," and GPO/FDLP, which would be responsible for "long-term access." This division mostly worked in the paper era when NARA was responsible for unique, unpublished Records, and FDLs were responsible for published

Public Information. It relied on the assumption that Records required an archive (NARA) to "preserve" them and published books needed libraries (FDLs) to provide "long-term access" to them.

But these assumptions had two fatal flaws in the digital age. First, agencies were neither required to send digitally published Public Information to GPO nor to schedule it to be preserved by NARA. This meant that, despite the assertions of the policy, neither NARA nor GPO could guarantee preservation of or access to digitally published Public Information. As we described in Chapter 5, published Public Information was falling through the cracks in the legal mandates for preservation. Second, digital publishing made the policy's paper-era distinction between "long-term access" and "permanent preservation" irrelevant. In the digital age, preservation and access are inseparable.

Digital information needs to be accessible for discovery and use, not just available as static containers (Chapter 17). Information that is not accessible is not preserved. In the paper era, the connection between access and preservation was, as Star and Ruhleder would say, "invisible" because the infrastructure ensured preservation though the distribution and retention of many copies in publicly accessible library collections. The reality was that libraries did "preserve" printed government information by providing "long-term access" to it. This fact became visible and a practical problem when the paper-era infrastructure broke.

Together, these mistakes and flaws created many of the gaps we describe in Chapter 13. This was a time that called for working toward a digital preservation infrastructure of many institutions, but GPO chose instead to rely on its own incomplete "storage facility."

Lasting Effects

One of the most important outcomes of the policy was the development of GPO's GOVINFO preservation repository and its certification as Trusted Digital Repository. This was a milestone and, as we write this, an achievement that is literally unique. Although GOVINFO is not an infrastructure, it is an important building block of what could become a Digital Preservation Infrastructure for government information. Neither the policy nor the law that enabled it required GPO to produce GOVINFO, but GPO's commitment to preservation (over a period of more than 14 years!) led to it doing so. This is a model of how strong infrastructures can be built in spite of the weaknesses of laws and policies.

But the policy's biggest and most lasting effect on the preservation of digital government information was to effectively replace more than 1.000 depository libraries with one underfunded agency. That was a mistake that GPO and FDLP have failed to acknowledge or fix.

FDLP libraries once played an essential role in making government information available to the public. When agencies made their published Public Information instantly available to the public on the web, that essential role was replaced by the web itself. But the role that libraries played in providing long-term access (i.e., preservation) through the paper era infrastructure was still needed but was not replaced.

Conclusion

GPO used paper-era assumptions to create a policy at the beginning of the digital era. That policy was convenient for the moment but neglected to look to the future and acknowledge the need for a new infrastructure.

Instead of working toward a new infrastructure and developing a role in preservation for its existing FDLP partners, GPO has doubled down on the policy's biggest failure. It has rescinded its requirement for selective depositories to select even a single "tangible" item and now promotes what it calls "All or Mostly Online Federal Depository Libraries" (GPO, 2018a; FDLP, 2022a). These "depositories" receive no digital deposits, build no digital government information collections, and preserve no born-digital information. They have no control over how digital information is indexed or discovered or delivered. The best they can do in this 30-year-old GPO model is provide a paper-era service of pointing with catalog records to the limited number of static digital objects GPO can acquire.

The shift to web publishing created a need for a new preservation infrastructure for digital government information. The quantity and variety of digital information produced by the government will require an infrastructure consisting of many partners. One agency funded to store information and without the authority to acquire information to store cannot do this alone.

Appendix C:
Selection

In this appendix, we briefly describe the concept of selection. This topic deserves its own book, but our purpose here is simply to introduce how digital selection differs from selecting paper records and publications. The way preservationists understand and implement selection will evolve as they gain experience with selecting digital information for preservation within the context of a Digital Preservation Infrastructure, which itself will evolve over time.

Two Kinds of Selection

The preservation of digital government information requires two different kinds of "selection." First, information must be selected for preservation. Ideally, this will create a comprehensive collection of preserved government information, managed as a single, universal collection that is freely available to all with centrally managed metadata that includes persistent identifiers (PIDs).

But such a collection will be enormous and difficult to use for any but the well trained. To make the preserved information truly accessible—not just available but discoverable, acquirable, and usable—to everyone, libraries will need to be able to select content and build sub-collections that are defined by the information needs of specific user communities.

We can refer to these two different kinds of selection as "building the comprehensive archive" and "building library collections."

Selecting to Build the Comprehensive Archive

To state the obvious, information will be lost if it is not selected for preservation. This means that an essential element of a Digital

Preservation Infrastructure (DPI) is the ability of the infrastructure to support comprehensive and precise selection. Digital preservation cannot be passive. In the paper era, papers could be forgotten for years, discovered in an attic, and rescued by preservationists. In the digital age, there is no guarantee that information that is not actively preserved will be findable later or rescuable if found. Information that needs to be preserved will have to be selected for preservation before it moves or changes or disappears (Johnson & Kubas, 2018).

The underlying challenge of preserving a comprehensive collection of government information is that it must include two very different kinds of information: unpublished Records and published Public Information (Chapters 3 and 17). Each will require its own approach to selection for building a complete archive.

Selecting Unpublished Records

Selecting Records for preservation in the paper era tended to be tied to records management. Records management practices typically designated most working records as temporary and without long-term value. As noted in Chapter 3, most unpublished government Records are not deemed worthy of preservation, and NARA preserves only 1 to 3 percent of them (NARA, 2020b).

Of course, the shift to digital record keeping may change how agencies and archives select records for preservation. In some cases, it may prove easier (or cheaper) to store digital records than to distinguish between those that should be discarded and those that warrant preservation. As it becomes easier to store records of the decision-making activities of an agency, it may become more desirable to preserve them. On the other hand, the costs of providing long-term stewardship of huge quantities of internal records may deter preservation. It may be that standards for record-keeping systems will change to address such preservation issues. Preservation could become easier and less expensive if agencies choose to adopt records-keeping systems that make records easier to preserve.

Nevertheless, in terms of selecting which unpublished digital government Records to preserve, the general procedures for creating Record Schedules that are approved by NARA and assigning NARA the task of preserving those records continues to make sense in the digital age. Unpublished Records are essentially unique and will require archival preservation and, unlike published Public Information, many of those unpublished Records will require review before being made available to the public. NARA already has regulations in place, and it set July 1, 2024 as the date it will stop accepting paper records from agencies (Musurlian, 2024; NARA, 2024).

Selecting Published Public Information

Selecting published Public Information for preservation presents a completely different set of challenges because of its defining characteristics. As described in Chapter 1, two of the justifications for preserving government information are its inherent value as a record of the actions of a government and the value of the information itself, which is often irreplaceable. As described in Chapter 5, published Public Information possesses, by definition, both of these characteristics. When an agency chooses to make information available to the public, it is explicitly identifying that information as having public value. And making the information available to the public is itself an official action, which is documented by the published digital object itself.

Indeed, although the practical criteria for preserving published Public Information and unpublished Records were different in the paper era (see Chapter 5), the law does begin with a presumption of preservation. Rather than asking agencies which Records should be preserved, it requires agencies to select which records to discard (44 USC 3303).

At first glance, this might suggest that the best strategy is just to preserve all published Public Information and not bother with the expensive and time-consuming task of selection. Indeed, some technologists have suggested just that, arguing that, since "storage is cheap," we should store everything and sort it out later. Preservationists, however, have soundly rejected this strategy because, although storage may be cheap, preservation is expensive (National Digital Information Infrastructure and Preservation Program [US], 2002; NARA, 2010; Tallman, 2021). The historian Abby Smith Rumsey belittled the strategy of trying to save everything as the "what-me-worry, storage is cheap, Alfred E. Neuman school of preservation" (Rumsey, 2011).

There are also practical barriers to preserving "everything." As we described in Chapters 6, 9, and 12, web harvests with very similar broad scopes and intentions have harvested very different content, suggesting that "everything" is an impractical goal for preservation—at least using current web-harvesting technologies.

The simple fact is that, although all published information has some value when published, not all published information has long-term value. It will be necessary to select information for preservation based on its long-term value, and that long-term value will vary for different communities of users. While some categories of information (such as Census data, laws and regulations, judicial decisions) have obvious and unquestionable intrinsic value, the value of other information may only be obvious to much smaller, more specific user communities. Thus, the value of information is tied to who uses it. A technical report on aquifers, for example, may have no interest to the general public but may be indispensable to geologists and farmers. Information that is critical to physicians may be of little interest to non-physicians.

Selecting information for preservation is essential in ensuring transparency and accountability in a democracy. This leads to choices that may be surprising. Consider, for example, information that is "out of date" or obsolete or explicitly superseded by other information. At first glance, it might appear that such information should definitely *not* be preserved! Indeed, NARA's *Guidance on Managing Web Records* says explicitly that "web content" should be destroyed "when superseded, obsolete, or no longer needed for the conduct of agency business" (NARA, 2005d). But while such guidance may be appropriate for unpublished information, it is often not appropriate for published information. The reason is that, to make government transparent and accountable, it is necessary to document the changes to government information.

In the normal course of business, governments often revise published information. When a new president is inaugurated, for example, one expects new policies. The number of changes and the speed of change may vary for different administrations, but we expect that every administration will be different in some ways from its predecessor. After all, that is part of the reason we have elections. Other kinds of government information are updated all the time, not just when administrations change. Laws and regulations are added and amended and rescinded, new economic and environmental and census data are collected and published, and government recommendations to the public (like the Department of Agriculture's "food pyramid" guidance) are revised. Changes in government information are normal in a democracy.

Because revisions to information is a normal part of government business, it is essential to preserve government information—even "non-current" and "out of date" information—in order to document those revisions. Citizens need a record of what a government's stated values were and when they changed, what actions it took and when it took them and which people were responsible for the actions, what data it collected and generated at specific points in time, and so forth. It is important to preserve even information that later proves to be inaccurate in order to document what the government knew and when it knew it.

Some digital publications, notably laws and regulations, have been designed to record changes. Other published Public Information, like census data, add new information rather than replace old information.

Other publications, particularly agency policy statements and communications and guidance to the public, are more ephemeral and, in the digital age, changes tend to replace, not supplement, previous versions. All these different kinds of published Public Information must be actively preserved or face almost certain permanent loss.

Trying to capture old versions of government information after they have been altered is at best inefficient and at worst impossible. A successful Digital Preservation Infrastructure will be able to preserve government information as it is published, before it is altered, erased, or superseded.

Selection of published digital information must also acknowledge the technical utility of the information and how it is used. Digital information is sometimes made available in more than one format or version, and different versions of the same content often have significantly different digital functionality. This will give each version a different use-value to different communities of users. A census micro-data datafile, for example, may be unintelligible to a high school student looking for the population of a city but indispensable to a graduate student in demography. Selecting "content" for one community does not necessarily satisfy the use-needs of a different community for the same "content."

Thus, selection of published information for preservation will be most precise and most comprehensive when it is driven by the information needs of specific communities of users. This suggests the need for many selectors, each applying the interests of a specific community to the selection process. This is significantly different than the traditional approach to preserving unpublished information, which centralizes the authority and responsibility for selection with few selectors (the agencies that produce the information under the guidance and approval of NARA) who are not end-users of the information. It is much closer to the FDLP model in which individual libraries select publications for their local communities.

As described in Chapter 21, a Digital Preservation Infrastructure that supports many different ingest streams can provide the flexibility to support the needs of even small communities of users. This can be more effective and more efficient than a centralized agency-based selection approach and can more realistically aspire to building a collection that is both precise and comprehensive.

Selection Criteria

It is difficult to know what will be valuable to users in the future. T.R. Schellenberg, the godfather of the appraisal of records, articulated this problem almost 70 years ago.

> In applying the test of importance, the archivist is in the realm of the imponderable, for who can say definitely if a given body of records is important, and for what purpose, and to whom? (Schellenberg, 1956)

The task is even more complex in the digital age because of the inherent characteristics of digital information. Preservationists must consider not just the "information content" (e.g., the words in a document), but also its presentation and its digital utility.

Presentation of information is a particularly complex problem of preservation. Consider as an example a simple Executive Order, number 13661 of March 16, 2014. It is available in at least eight versions:

- as an HTML file on the Federal Register website https://www.federalregister.gov/documents/2014/03/19/2014-06141/blocking-property-of-additional-persons-contributing-to-the-situation-in-ukraine (which notes that "the PDF linked in the document sidebar" is "the official electronic format")
- as that official PDF file as published in the *Federal Register* https://www.govinfo.gov/content/pkg/FR-2014-03-19/pdf/2014-06141.pdf

- as a plain-text version of the *Federal Register* version https://www.govinfo.gov/content/pkg/FR-2014-03-24/html/2014-06612.htm
- as a different HTML file on the White House website https://obamawhitehouse.archives.gov/the-press-office/2014/03/20/executive-order-blocking-property-additional-persons-contributing-situat
- as a PDF file as printed in the *Compilation of Presidential Documents* https://www.govinfo.gov/content/pkg/DCPD-201400171/pdf/DCPD-201400171.pdf
- as an HTML file of the *Compilation of Presidential Documents* version https://www.govinfo.gov/content/pkg/DCPD-201400171/html/DCPD-201400171.htm
- as a PDF file of the 2015 Annual Edition of the *Code of Federal Regulations*, Title 3 https://www.govinfo.gov/content/pkg/CFR-2015-title3-vol1/pdf/CFR-2015-title3-vol1.pdf
- as a plain-text version of the 2015 Annual Edition of the *Code of Federal Regulations*, Title 3 https://www.govinfo.gov/content/pkg/CFR-2015-title3-vol1/html/CFR-2015-title3-vol1.htm

Making the selection decision more complex is the fact that each of these presentations may appear differently to different users using different browsers or PDF readers on different computing devices. The appearance of content encoded in HTML is particularly unpredictable because some browsers allow users to choose to use "dark mode" or "reader mode" and allow users to block fonts and stylesheets and JavaScript from loading.

The question the preservationist must face is what matters to the community for which the information is being preserved. For example, is the Designated Community interested in the words in the document or the visual layout of the words (its presentation)? This leads to other questions such as how much information value is imparted by the layout? Do all versions have the same words? Is one version "official"? Is one version digitally encoded so that a user (or a library providing a digital service) can make its presentation more readable or understandable—even if that presentation is different from the original? Does one digital encoding of the information make the information more usable (e.g., for computational analysis) than another encoding? Because some of these different versions exist because they were included in traditional printed publication, there may be a value in ensuring those publications are completely preserved even if the information exists elsewhere.

In addressing these questions, selection decisions must be made in the context of the information needs of particular Designated Communities. Different communities may have different use-needs that each require different versions of the same content.

An approach that selects information for preservation based on the needs of a Designated Community might lead a preservationist to ask if their Designated Community is interested in:

1. the information value of the Public Information (the "content");
2. how the information is presented (e.g., the preservation value of a government produced comic book or consumer pamphlet may be driven more by its form of presentation than by its information content);
3. the way the digital information is packaged for use and re-use (e.g., one community may prefer a spreadsheet and another community may prefer a PDF of the same information content).

Each of these three reasons for preserving information presents its own preservation challenge, but preserving the information content itself is perhaps the most perplexing challenge. It is difficult because some elements of the presentation of digital information impart meaning to the content. Appendix D describes this issue.

Selecting to Build Library Collections

Once information is successfully stored and managed and made available with persistent identifiers, libraries will be able to do more than point at selected, static documents. They will be able to build defined collections and develop digital services for those collections. They will be able to deliver collections and services that meet the needs of specific communities of users.

Why Build Library Collections?

As the DPI becomes more successful in preserving more information, the complete archive will grow in size. Dramatic improvements in selection and ingest will result in dramatic increases in the size of the archive. As the size of the archive increases the variety of file-types and formats and functionalities of the digital content preserved will also increase. The complexity of the metadata will increase. It will have to identify and connect individual digital objects, guarantee the authenticity of those objects, ensure their functionality, and provide their temporal context and provenance. Managing the complete collection will be a very big task, but it is only half the task of successful preservation. The other half is to make sure the content preserved is discoverable and usable by the people who need that information.

As the complete archive grows in size and complexity, it will become more difficult for users to find what they need. This is the needle in the haystack problem: the bigger the haystack, the harder it is to find the needle. The complete archive will, much like the govinfo.gov website does today, make it relatively straightforward to find known items (i.e., a specific publication with a particular title created by a specific agency at a specific point in time). Even an enormous archive with many types of information can support such discovery service using basic descriptive metadata, simple organization of information, and simple searching and display options for users.

But it is much more difficult for an enormous, diverse, comprehensive archive to provide adequate discovery tools to users who are not looking for a publication but for information. When users are looking in such an archive for a fact, a subject, or a type of information and do not know which agency or publication might have what they need, general searches are not adequate. (This is, presumably, why GPO defines its Designated Community for GOVINFO to be those who are "familiar with the organizations, documents, publications, and processes of the legislative, executive, and judicial branches of the United States Federal Government" [PTAB Primary Trustworthy Digital Repository Authorisation Body Ltd, 2018].)

When users are not looking for a publication but for information, they need a "library" more than an "archive." Where the traditional archive role is managing a comprehensive collection of information, the traditional library role is to make information more accessible and usable for people (Chapter 17). Providing collections and services (two of the main traditional roles of libraries) are roles that remains essential in a Digital Preservation Infrastructure. A successful DPI will need to make it easy for libraries to build defined digital collections that include selective parts of the comprehensive collection. This will involve libraries selecting content from the comprehensive collection.

These sub-collections will have three benefits for users. First, they will bring together into a searchable collection the information that meets the needs of a community. (From the user perspective, this changes the problem from looking for a needle in an enormous haystack to looking for a hammer in a hardware store.) Second, libraries will be able to build collections that combine government information with non-government information. Users will be able to easily navigate from journal articles that use government data to the government data used, and from government information to non-government information that references it. This will require non-government libraries whose missions are different from the missions of government agencies.

Third, the library that builds a defined collection will be able to develop digital services customized for that collection. A legal collection might bring together information spanning the creation and instantiation of laws, the regulations and administrative enforcement of those laws, and the judicial review and enforcement of those laws with indexes that point to and link to information rather than to the publications that contain that information. A collection of survey data might provide the ability for users to search for survey questions asked, populations surveyed, and time periods covered and be able to deliver results as statistical tables or data ready for use by the user's preferred statistical software.

The possibilities for building collections and digital services are unlimited, but only a finite number of collections will need to be built to provide adequate services to users. Libraries that participate in the Digital Preservation Infrastructure will be able to identify gaps in the existing mix of collections and services based on the needs of their own communities of users and develop services that target those gaps. As more libraries participate, there will be fewer gaps and more opportunities for existing services and collections to overlap and coordinate into hybrid collections. A preservation plan that anticipates this can facilitate it by providing technical standards for sharing and by managing complex metadata. As digital services proliferate, there will even be opportunities for the development of digital services that help users locate the best digital service for their research.

Libraries will benefit from building defined collections in several ways. They will fulfill their missions directly by providing collections and services that they have identified as being critical to their own communities. By participating in the Digital Preservation Infrastructure, they will make the infrastructure itself more robust and valuable and will make digital services developed by other libraries available to their communities. And individual libraries will become nationally known for their collections and services, which will be available to all. Just as in the paper era when libraries defined their identities and their value by their collections, so it will be in the digital age.

A system of Primary Archival Repositories, such as we describe in Chapter 21, and the infrastructure itself will benefit from the existence of defined collections as well because these sub-collections will be integrated with the comprehensive collection. Collections will not be "silos" of information that are separated from related information. Rather, they will be connected, through metadata, to their original sources in the comprehensive collection and to other sub-collections. This will make it easy for users to start with one collection and easily navigate to other collections with related information. The infrastructure itself will grow stronger and more useful as collections grow.

How to Build Collections

Even before the emergence of a Digital Preservation Infrastructure and its specific strategies, standards, and technologies, it is easy to envision at least three ways that libraries will be able to build digital collections of preserved government information. All three will make use of preserved digital objects and their metadata that the comprehensive archive has but will differ in how they use these components.

1. Use existing metadata that points to existing digital objects in the comprehensive collection but organize and index the metadata differently and present it with a user interface that is customized for a specific Designated Community.

2. Use existing metadata that points to existing digital objects but obtain or create new metadata for those objects in order to organize and index the data in a new ways and provide a user interface that is customized for a Designated Community.

3. Obtain copies of selected digital objects from the comprehensive archive, create and host new digital objects from them, and create new metadata for the new objects in order to create a customized user interface to this new collection.

Each of these tactics will use metadata both to identify the information in the collection and to create an information service for that information. (A successful DPI will preserve and manage metadata for use and reuse by all participants. Appendix E describes some of the characteristics of metadata for a DPI.) At its simplest, libraries could use existing library catalog technologies as a user interface to DPI metadata.

Nevertheless, in order to build full-text indexes and customized and robust interfaces to complex metadata, libraries may also choose to use specialized software and develop customized databases with specialized indexes. Libraries may also wish to transform preserved content itself in order to enhance its functionality for the end user. The possibilities are limited only by the imagination of libraries seeking to meet defined needs of their communities.

Appendix D:
Significant Properties

In Chapter 2, we defined information as the stuff we can preserve, whether a book or a bitstream. But all stuff is not the same. Printed books include a fixed presentation—an unalterable "user interface" of the communication being preserved; bitstreams do not. And bitstreams can be used in ways that ink on paper cannot.

This creates an issue for digital preservation. We use books directly, but we use bitstreams indirectly, mediated by computers. Andrew Wilson of the UK Arts and Humanities Data Service and the National Archives of Australia describes this mediation this way:

> The data source (record, object, etc.) needs to interact with a process in order to be understood by a user. That process is a combination of hardware (computer) and software (application + operating system). Without this mediation the digital object is meaningless since the data that makes it up exists independently only as a stream of bits. (Wilson, 2007)

The simple fact that "the digital object is meaningless" without mediation is a primary problem that libraries and archives understand well. But it is more than being able to make a bitstream "technologically viable" (Barnum & Kerchoff, 2000). Bitstreams can have many properties that need mediation and not every preservation tactic is equally successful in mediating all properties. This can lead to a preservation challenge of determining which properties are essential and which are not.

Properties may be explicitly described inside the bitstream of a digital object (for example, HTML markup in a web page), or they may be explicitly described in a separate metadata bitstream (for example, a "codebook" that describes the structure of a datafile), or they may be implicit and not exist until computed at the time of presentation (for example, the number of pages in a document). Christoph Becker of the Faculty of Information at the University of Toronto describes properties that are computed during mediation as "emergent properties." All such properties, whether explicit or implicit, must be mediated to be used by humans. Preserving the bitstream only goes halfway. Ensuring the ability to computationally mediate the bitstream at the time of use (or as Becker calls it, the time of "performance") is the other half of preservation (Becker, 2018).

The digital preservation literature uses terms such as characteristics, attributes, and elements to refer to such properties. Regardless of the term used, some, but not necessarily all, may be considered "significant" or "essential" (Wilson, 2007). The National Archives of Australia prefers the almost poetic term "essence" to describe "the characteristics that must be preserved for the record to maintain its meaning over time" (Heslop et al., 2002). Ross Harvey and Jaye Weatherburn in their textbook on preserving digital materials put it this way:

> [W]e need to know more about exactly what it is we are
> trying to preserve. What characteristics, attributes, essential
> elements, significant properties—many terms are used—of
> digital materials do we seek to retain access to? What is the
> "essence" of digital materials? Whereas this question was a
> simple one to answer in the non-digital context, where
> typically we sought to conserve and preserve the original
> artifact, for digital materials it is not so straightforward.
> (Harvey & Weatherburn, 2018)

The issue of which properties are significant is not a new problem. It was mentioned in one of the earliest reports on digital preservation (Waters & Garrett, 1996) and has been discussed and analyzed ever since. The discussion continues because there is no single, simple way to define which properties are significant. Each archive must determine for itself which properties are significant. An RLG/OCLC report described it this way:

> A digital object's significant properties are not absolute, nor are they static. How a repository determines which properties are significant depends on the nature of the organization, the services it provides, and its role in preservation. (Research Libraries Group, 2001)

These comments define the problem nicely. "We need to know more about exactly what it is we are trying to preserve" and "which properties are significant depends on the nature of the organization [and] the services it provides." This means that, in order to choose which properties are significant enough to preserve, preservationists must know which properties are significant for the community for which they are preserving the information. Preservationists must know how that community of users intends to use the preserved information. The properties of any given digital object that are significant for one community may be irrelevant to another community.

Digital preservation is so much more than storing bits. It must also preserve the significant properties of those bits. Determining which properties are significant is complex because properties can affect both the presentation of information and its usability.

Presentation

The presentation of information is an essential element of communication. Some properties of the presentation of words and images in a publication convey explicit meaning and are essential to making the information understandable. "Information" is not just a "bag of words"; it is text and images ordered and presented in a specific way (Joulin et al., 2016; Jacobs & Jacobs, 2016). A table of population statistics conveys meaning by the placement of numbers in specific columns and rows. The size and position of text provide semantic context that indicates which text is a chapter heading and which is a page number. Colors, fonts, locations on the page, boxes, lines, and other graphic elements give context to the text (e.g., callouts, sidebars, illustration labels, annotations, footnotes, epigraphs, headings, and so forth). Loss of any of this presentation information can result in loss of meaning (Jacobs & Jacobs, 2013).

Other design and presentation decisions do *not* convey specific information but may affect how readers react to or interpret the meaning (Bartram, 2001). The design of a document may, for example, suggest its intended audience, e.g., a scientist or a consumer, an adult or a child. And subtle design choices may affect how readers use or interpret the meaning. An animated graph may be more convincing than a simple table of numbers though both contain the same content. Visual designs and layouts also provide "branding" that distinguishes publications. Such properties can affect how we interpret and remember information (Douneva & Jaron, 2016). Maybe we shouldn't judge a book by its cover, but we often do.

Although digital presentation shares the functions of print presentation (affecting how clearly the meaning is conveyed, how easy it is to use, and how the user responds to it), digital information properties differ from print in several ways that affect its preservation. Digital publishing offers authors and designers new ways of presenting information that are not available to printed publications. These include dynamic content, animation, audio/visual elements, and, of course, hypertext for the creation of non-linear arrangement of content (Rowberry, 2023). Often, a single bitstream (such as a web page coded in HTML) must draw in other bit streams (images, stylesheets, JavaScript code, etc.) to be complete. Some of these various presentation elements are essential to the meaning and functionality of the information being preserved. Other elements can enhance the user experience and the communication of ideas and facts and data and opinions. Some, however, may be irrelevant to the "essence" of the information.

But the digital presentation of information also has drawbacks that print generally does not. When design of a website (for example) is given to developers who have no experience with conveying government information to the public, the design may include features that are unnecessary and even intrusive. The technology of web development can have a profound effect on the appearance of information. Anyone who has browsed different versions of a website using the Wayback Machine has seen the often dramatic, usually inessential, changes to the presentation and appearance of the site over time. Primitive HTML-coded-by-hand websites have been replaced with commercial styles delivered by commercial Content Management Systems. These have definitely changed the appearance of websites, but it is not always clear that they have enhanced the communication of information in any useful way. The fact that the same textual government information is sometimes available in different presentations by different agencies (see Appendix C) is another example of the complexity of determining what is significant and what is not.

Design elements can affect the usability of a digital document negatively as well as positively. Today's web design choices, for example, often make it difficult just to read text. Presentation of information on the web has become so bloated with unnecessary and even intrusive "features" that web browsers have introduced a "reader mode" that removes "un-useful page elements" and gives the user control over font choice and size in order to make the text easier to read (Ghasemisharif et al., 2019; Burke, 2023).

This highlights the fact that the presentation properties defined by the publisher will be interpreted by various user-chosen mediation tools (e.g., a device and its screen size, its operating system and software, and the settings of the software). The task of sorting out which presentation properties are significant enough to be preserved must acknowledge that those properties may not predict how the bitstream will appear to any given reader. Two people using different copies of the same book are given precisely the same singular presentation of words and images as designed by the publisher with the intent of communicating particular knowledge. This is true even if the readers are decades apart. But two people loading the same web page today may see a completely different presentation of the same content.

All too often, web design obscures information content by presenting information that is accessible only visually, which makes it unavailable to the screen readers that visually impaired users rely on (Borodin et al., 2010). Those who create digital documents can also neglect (sometimes intentionally) properties that make the information machine-actionable. Two common examples of this are PDF documents that are simply images of printed pages, lacking the functionality to search or copy text, and PDF files with numeric tables that cannot be copied for analysis, re-use, or repurposing.

The preservationist must face the existence of presentation properties of digital objects and the ability of mediation tactics to accurately reproduce those properties in the future. When the preservationist cannot guarantee accurate mediation of all elements, a simple question arises: What are the essential elements? One can imagine any number of such specific examples. Does the typeface and size of font need to be preserved? How about the color of the background and the color of the font? If a web page has navigation tools for the website, must those be saved along with the "content" of the web page? If a web page has dynamic content such as the current temperature should the temperature at the time of capture be preserved, or should preservation capture the ability to display the current temperature whenever the preserved page is viewed?

These questions cannot be answered definitively in the abstract. The answers depend on who will be using the information and how they will use it. In other words, the answers depend on the Designated Community.

Preservationists may also wish to be able to repurpose and re-present information in new ways. Simply applying a new stylesheet to an HTML web page can make it easier to read and understand. Dividing long documents into its constituent parts for better indexing and delivery can enhance the utility of digital information.

Although the questions and issues that presentation properties raise may seem problematic, they also can suggest approaches to solutions, which we describe below.

Utility

The problem of preserving essential properties is more than preserving presentation and appearance. Digital publications may also be machine-actionable. Any property that allows a user to actively interact with a digital object is a machine-actionable property. This can include simple functions that we take for granted such as being able to search a text or copy text or images or modify a document with annotations and highlights and links. It can also include more elaborate interactions in which a user points a computer at a publication and tells the computer how to interact with the digital object. Such interactions are the primary purpose of some digital objects (such as databases, spreadsheets, geographic information, etc.). But, increasingly, researchers use technical means to interact with publications that were designed as static documents intended primarily for reading directly. Techniques such as "distant reading" and the computational analysis of whole corpora of documents have become common.

The preservationist will need to acknowledge and preserve those properties that enable future users to interact with digital objects in the ways they wish—even if those were not the originally intended uses of the information.

Solutions

Although there are no definitive one-size-fits-all solutions for identifying and preserving the "essence" of digital information, there are approaches that can address the needs of users. Different approaches will be necessary as digital publishing evolves from its current practices (which often emulate print publishing) to new ways of publishing and distributing and consuming information based on the inherent characteristics of digital information.

Current Approach

The current approach to preserving the significant properties of digital objects is to rely on technology. This approach addresses the significant properties of digital objects indirectly by focusing on the most widely discussed preservation problem: how to ensure that we can represent a preserved bitstream *at all* in the future. The most common technical tactics for this are migration and emulation (Cornell University Library, 2003).

To oversimplify, migration converts the bitstream to make it work with changing technologies and emulation leaves the bitstream alone and attempts to preserve (or emulate) the original technology under which the bitstream works (Heslop et al., 2002). Ross Harvey and Jaye Weatherburn call these techniques "preserving technology" and "preserving objects" and point out that they are two ends of a spectrum of techniques (Harvey & Weatherburn, 2018).

Although any specific tactic can be successful in some cases, none can guarantee success for every bitstream or every property or every use-case. Shortcomings and failures of these tactics may not become evident until people in the future attempt to use them. This conundrum is particularly problematic when a tactic is chosen as a one-size-fits-all, universal solution. One-size-fits-all solutions work best when most digital publishing attempts to replicate printed publications. In that instance, the goals of the publisher and the reader are quite similar and the objects of preservation are static and well understood.

But as digital publishing evolves past the print-era model of static publications, the properties of digital objects become more complex, and preservation becomes more difficult. One can, for example, easily imagine a tactic that successfully displays pages of a PDF file but fails to open files of other formats embedded in the PDF file (Lazorchak, 2012; Wyatt, 2024).

The long-term costs of relying on migration and emulation are unpredictable and, though either may be potentially effective technically, it is not clear whether either will be affordable (and therefore practical) in any particular case.

Interim to Long-Term Approach

Choosing a tactic (such as web harvesting and web display) that focuses on the bitstream, and its mediation will become less successful as digital publishing evolves. As more information is delivered from databases, it will become increasingly difficult to harvest. As new digital formats are created for digital-only consumption, mediation will become more complex and expensive.

A different approach would be to choose a preservation strategy that is based explicitly on the information needs of people who want the information preserved. This will require a digital preservation infrastructure. As such an infrastructure emerges, it will be possible for preservationists to start developing tactics that focus on the needs of specific communities of information users. These tactics will focus on the significant content of information objects, its significant properties, and its usability rather than on preserving either the technology or the digital objects. In the medium term (as digital publication evolves), this may mean putting less emphasis on preserving the original bitstream and more emphasis on preserving the content and its usability even if that means altering the original bitstream. Such an approach would allow an archive to tailor its preservation and mediation tactics to ensure that that a particular Designated Community can use the information in the way it prefers. The properties that the community needs would drive the technological tactics of preservation.

Focusing on users and use could mean that the same content might be preserved more than once in order to address the different needs of different communities. To use a simple example, census data might be preserved in two ways. One way would preserve a particular format and mediation technique that would presume its use would be by those people seeking to look up facts and display tables of statistics. A separate way would preserve the same content for researchers who wish to use statistical software to analyze the data. By addressing the significant properties that a community requires, an archive can make a user-oriented selection decision and ensure that their Designated Community will not only have the content they need but will be also able to use the properties of that content in their preferred fashion.

This suggests the need for a focus on significant properties at the time of ingesting content. As an example, take a website that provides access to a bibliographic database. One Designated Community might view the "look and feel" of the website as its most significant property, and another might view the bibliographic records in the database as the most significant property of the website. One archive might use web-harvesting software to capture the website's presentation of records and OpenWayback or similar software to recreate the way the website functioned. Although, as we described in Chapter 6, this might fail to gather all records in the database, it might be the right one for this use-case. The other archive might acquire a copy of the raw data in the database and then develop a digital service to provide users with access to the data. This would require the archive to develop a way to acquire the raw data. (In the example we used in Chapter 6, the data was available in GitHub.) It would also require the archive to develop a digital service to provide users access to the bibliographic records. Although these new tasks are not as easy as using the one-size-fits-all web-harvesting approach, they have short-term and long-term benefits. In the short term, the data are preserved and the immediate needs of the DC are met. In the long term, the archive gains the flexibility to adapt and evolve its digital service to meet new needs of new users in the future.

Census data provides a similar example. Because the Census Bureau no longer provides print-era style publications (printed volumes of statistics or PDF facsimiles of printed volumes), there are no "publications" or "documents" to preserve. The bureau has replaced providing the same publications to all users with providing a service to each user. Rather than trying to duplicate an existing digital service that delivers statistical tables in response to queries, preservationists could choose to preserve the data behind the existing Census Bureau service and develop their own digital services to deliver that content to users.

Developing such digital services will be a big task, but it is likely to be the future of the digital archiving. Increasingly, information is stored in databases. Even when that information is delivered as "documents" or "books," it is likely that the information is stored in a database and will be converted from a neutral format (such as SGML or XML) into a user-preferred format (such as PDF or ePub or responsive HTML) at the time of request and delivery. Attempting to preserve each format would be difficult and expensive and prone to error. Preserving the raw data in the neutral format and providing a digital service for delivery of the information would not only be less expensive in the long run, it would also, as noted above, enable the archive to adapt the service over time to meet the evolving needs of users in the future. This approach focuses the archive's actions on preserving the significant properties of the targeted information. Preserving databases will require new approaches to selection and acquisition; web harvesting will rarely be able to reach the databases behind digital services. New approaches could include use of agency-provided APIs and formal or informal agreements with agencies. As the DPI evolves and libraries are able to demonstrate their value in providing sophisticated digital services, some agencies may seek out partnerships with libraries to preserve their information. Developing such new approaches will be both more effective and more efficient in the long run.

In our suggested approach to a Digital Preservation Infrastructure (Chapter 21), Ingest Repositories would each focus on the use-needs or use-cases of specific Designated Communities as a selection strategy. This would ensure that the comprehensive digital archive would include different kinds of content (when needed) and ensure that each was preserved using the techniques necessary to ensure its future use by people. In each case, the specific properties of the information would be preserved for the community that found those properties essential. Libraries that provide Defined Collections and Services would be able to specialize in types of content and types of services and types of delivery. By working within the context of a universal preservation plan and conforming to standards developed by the planning and implementation committees, there would be no need for duplication of services because all users will be able to use all defined collections and services.

This new, medium-term approach to ensuring that the significant properties of digital information are preserved and usable focuses first on the needs of Designated Communities and then selects and ingests information in such a way that meets their long-term needs. This approach could continue to function in the long term as digital publishing evolves.

Long Term Approach

The approaches described above address digital information as the government produces it today. They are mechanisms for coping with literally hundreds of file formats that are posted on the web always with the intent of being read or used immediately and not necessarily designed for being used in the distant future. This is the information landscape of the mid-2020s. But digital publishing is evolving. In the long term, more information can be designed for long-term access and preservation.

To ensure that the "essence" of digital information is preserved, preservationists need to participate in this transition. The long-term approach is to participate in the design of a future digital publishing infrastructure that will be built for preservation. For preservationists to participate in the transition, it is necessary to have an understanding of digital publishing now and where we want it to go in the future.

Digital Publishing Now

During this early stage of the transition, most government information is simply a digital replica of printed publications. These make little use of many of the inherent characteristics of digital information, but paradoxically, they can be difficult to preserve because they are rarely designed with long-term access and preservation in mind.

The reliance on the PDF format for distributing government information is evidence of the influence of the print era on the current distribution of digital government information. As noted in Chapter 6, PDF is one of the most popular formats for publishing government information, and the government (apparently) publishes millions of them each year. The PDF format was designed to make page-oriented documents that could be viewed and printed (Warnock, 1995). The PDF technical specification describes the PDF format as representing "electronic documents," which echoes the 1968 definition of government publications as information "that is published as an individual document." GPO says in its publishing guidelines that the PDF format is "[t]he most common format for presenting documents online" largely because PDF files "maintain product design integrity (e.g., page formatting)" (GPO, 2015b). As GPO Director Hugh Halpern has noted, some of those product designs "have remained relatively unchanged for the last 150 years" and were designed with properties such as small type sizes and tight line spacing, "for economy of printing, rather than for readability" (GPO, 2020a). Those are embedded properties that probably few users would judge to be significant for understanding or using the content.

Another bit of evidence of the government's reliance on print-era models of information distribution is the continued publishing of print-like books such as hearings and print-like serials such as the *Federal*

Register. These retain all the trappings of their print counterparts: titles, page numbers, page layout, serial issue numbers, and so forth.

Relying on the print model for digital publishing has some advantages. It has tradition and familiarity to its credit, and it makes it easy for users of the information to refer to and cite publications accurately. It also reflects the need for information that is "fixed," not "fluid" (Levy, 1994). Most government information is intended to be fixed and stable over time. For governments to be accountable to the people, there must be a fixed record of government actions. Emulating the paper model of publishing provides a way to emulate fixed text. And most government information is text. As we noted in Chapter 6, although there is a significant amount of born-digital content being produced by the government in non-text formats (e.g., audio files, video files, spreadsheets), the vast majority of content is still text-based.

Laws and regulations provide a prime example of the importance of texts remaining stable. These have to be fixed so that they can be enforced. But even laws and regulations change from time to time. The paper era defined elaborate mechanisms for recording and documenting these changes. This enables anyone to be able to see the current text but also to track previous versions and how and when they were changed.

But the guarantees of print-era models come with risks in the digital age. One expects many other government documents (such as congressional hearings, agency annual reports, technical reports, judicial decisions, and so forth) to remain unchanged after their publication. In the print era, there was no need for mechanisms for tracking changes to such publications since they rarely if ever changed. But in the digital age, it is technically possible—even easy—to alter such publications. When users must rely on the copy that the producer makes available on the web, there is the implied risk of the producer making changes without notifying the public or retaining the old version for reference. Lacking procedures for recording and documenting changes, there is the strong potential for losing information and for having no way of recognizing or documenting the loss.

Still other government information, particularly web pages that agencies use for day-to-day communication with the public, carries little or no expectation of stability. A page announcing meetings and agendas changes as often as the meetings and agendas change. Documents that communicate or explain agency actions and policies will almost certainly change when administrations change and introduce new policies or modify old ones. With information being created and distributed digitally, changes in policy could result in old, "out of date" documents being altered or deleted without notice. Without a robust digital preservation infrastructure to guarantee the survival of those old policy documents, new policies could effectively not just change old policies, but erase them from history. The same could potentially be done to irreplaceable data collected and compiled by the government (Dennis, 2016; Malone, 2016; Bergman & Rinberg, 2018; Gehrke et al., 2021).

During this period of transition between the paper era and the digital age, the one way to ensure that information is not lost is to preserve information as it is released and published. This fits well with the short-term approach described above.

The Future of Digital Publishing

Digital publishing is evolving. The long-term approach to addressing the significant properties of government information is to participate in designing the next digital publishing infrastructure. Preservationists can ensure that the goal of digital publishing in the future will be preservation of information, not the size of a printed page.

A lot of progress has already been made, particularly in the area of legal information. We are already seeing new publication formats that are not just modernized or innovative, but transformative. Paper-era models for publishing legislative bills and laws and codes already had a highly structured format (House Office of the Legislative Counsel [US], 2023) that can relatively easily be translated into digital citations and persistent URLs (Hershowitz, 2023). Congress, the Library of Congress, GPO, and GAO have been cooperating since 2012 on the Congressional Data Task Force (formerly the Bulk Data Task Force) to build new standards for legislative digital publishing. The task force has defined and promulgated the use of the United States Legislative Markup (USLM) schema, an XML standard that describes both the presentation and structure of a document (US House of Representatives, Office of the Law Revision Counsel, 2016).

USLM is already being used for publishing bills, resolutions, statutes, and the US Code. It is designed to be flexible and adaptable to other legislative materials and to meet future needs. GPO's new "XPub Program" (formerly known as GPO's "Composition System Replacement") (GPO, 2025b; Clerk of the House [US], 2024), is designed to work with USLM and deliver content to users in a variety of formats including responsive HTML that displays properly on all devices. This is a working example of the idea described above of storing (preserving) information in "neutral formats" and delivering it in the format preferred by the user. GPO already has other projects in the works to use USLM, including producing the *Constitution Annotated* and "one-off" reports (GPO, 2022b).

This is an important development in the long-term approach to preserving the essential properties of digital government information. The first generation of USLM was only designed to express the visual characteristics of a document. But its second generation, introduced in 2013, goes beyond describing the "presentation" of information to describing the structure and meaning of information (Legislative Branch Innovation Hub, 2019). This new standard takes the guesswork out of understanding the essential properties of a digital government information. It is explicitly designed to "[p]rovide a flexible foundation to meet future needs" (US House of Representatives, Office of the Law Revision Counsel, 2016). The evolution of USLM demonstrates how standards like this will be able to change to meet new challenges (McLaughlin & Stover, 2021).

GPO offers information in XML formats for bulk download in order to "maximize the ways this data can be used or repurposed by users" (GPO, 2016a). In addition to USLM, GPO also offers the *Code of Federal Regulations* in an XML format (GPO, 2015c).

Clifford Lynch described the process of defining the significant properties of digital objects in the objects themselves as "canonicalization" (Lynch, 1999). He suggested that canonicalization could help "to make precise what is important about a class of objects." He suggested that authors of digital information objects should be able to "make implied statements about what they view as the minimal, essential version of the work." Defining "a canonical form for a class of digital objects" would enable us to capture "the essential characteristics" of a digital object. It would do more than provide a solution to the significant properties problem. "[C]anonical formats and canonicalization algorithms (that is, algorithms that compute canonical representations) for various types of digital objects" could address many of the problems of migrating formats including authenticity, provenance, and integrity. He noted that some types of digital objects, such as XML, images, and sound files, already have a logical and fairly straightforward set of minimal characteristics. While it is more difficult to define the minimal essential characteristics of text documents, the example of legislative documents and the USLM format suggests that it is already being done.

Defining a digital publishing infrastructure may sound like an enormous task, and it is. Many communities of producers and users of digital information will hold a stake in the development of such standards. If preservation is to be built into that standard, preservationists must participate in that process. This is yet another role for individual librarians and archivists and communities of librarians and archivists in a Digital Preservation Infrastructure (Chapter 22).

Appendix E:
Metadata

A successful Digital Preservation Infrastructure (DPI) will need to support extracting, creating, sharing, managing, and using large volumes of metadata. In an era of superabundance of information, Defined Collections will be even more important than physical collections were in the paper era, and they will be defined not by the location of objects in a building or at a particular repository or on a particular server, but by metadata. Digital services for these digital collections will depend on rich, accurate metadata for discovery and delivery of content so that it can be shared, used, reused, analyzed, and repurposed.

In order to accomplish all this, preservationists need to have a clear conception of an infrastructure for metadata that supports a DPI. This will expand and transform the print-era model of one single, multi-purpose descriptive record for each publication.

The design and construction of a metadata infrastructure for a DPI will require a great deal of discussion and experimentation and development. In this appendix, we intend only to introduce the topic to those who are not metadata professionals and particularly to digital preservationists who are thinking about the future of metadata as part of a DPI.

Data or Metadata?

Literally, metadata is data about data. It is information that defines and describes other information (Gilliland, 2016; ISO, 2023). Metadata does not have to be digital. Card catalogs in libraries were metadata. But an infrastructure for digital information will use digital metadata, and digital metadata is also digital data.

Digital metadata can be used in more than one way. Take, for example, the MARC (Machine Readable Cataloging) metadata records that libraries have long used to describe their holdings. When a million MARC records are loaded into a library's OPAC (Online Public Access Catalog) and presented as a tool for finding digital objects, those million MARC records are being used as metadata for those million individual items. When those same million MARC records are analyzed to examine the quality and accuracy of cataloging, those records are being used as data in their own right (Mayernik, 2011; Draper & Lederer, 2013; Wahid et al., 2018).

Similarly, information that was not originally designed as metadata can be used as metadata. For example, the text of a document can be used as metadata by incorporating it into a full-text index to a collection of documents. It is all information. "Data" can be used as "metadata" and "metadata" can be used as "data." As the International Organization for Standardization (ISO) says (ISO, 2023), "[D]ata become metadata when they are used in this way."

A successful Digital Preservation Infrastructure will build on the potential flexibility of metadata by storing information in ways that will allow it to be used and re-used in different ways for different purposes—including, most importantly, *future* uses that we cannot anticipate today. An understanding of the components of metadata and the written standards that define those components will be essential.

Components of Metadata

Three components are common in all useful metadata. These components are typically defined in a published "standard" or "schema," of which there are many.

- **Elements**. Metadata standards begin by choosing the properties of information ("elements") that are most important for a particular use. Elements of a book might include title and author, and elements of an image might include color palette and

365

dimensions. A standard will identify, name, and describe elements that serve a purpose. Libraries use the *Anglo-American Cataloging Rules* (AACR2), for example, to describe books. Digital preservation repositories often use a standard designed specifically for preservation metadata (PREMIS), which defines elements that enable the repository to ensure that a digital object remains viable and renderable and unaltered.

- **Value Rules**. For every object that metadata describes, the metadata will record specific values or "attributes" for each of its elements. A "title" element, for example, might have the value "The Sun Also Rises." The standard usually imposes rules that define how those values must be expressed and may specify the use of authority lists such as ORCID (the Open Researcher and Contributor ID), which uniquely identifies authors of scholarly communications, or ontologies such as the LCSH (*Library of Congress Subject Headings*). In general, rules can help ensure that a metadata standard is implemented consistently and that the metadata are unambiguous.
- **Encoding**. A digital metadata standard may include specific rules for how metadata must be encoded into a specific kind of "file" or bitstream. MARC is such an encoding, but modern metadata standards tend to use XML and more open encoding standards such as the Resource Description Framework (RDF).

Standards

There are hundreds and potentially thousands of metadata standards (Greenberg, 2018). There are so many standards because there are so many potential uses of digital information and so many ways of describing, identifying, and using digital information.

Some standards are designed for specific functions such as preservation or description. Others are designed for specific domains or disciplines. (The *Metadata Standards Catalog*, for example, lists more than 150 subject areas [Research Data Alliance, 2024].) Others are designed for specific types of objects that the metadata describe, such as music, text, data sets, or images. When no existing standard works for a particular community or function, there are "application profiles" that modify the standard (Tambouris et al., 2007). There is, for example, an application profile of the Dublin Core metadata standard that is designed specifically for government information (DCMI-Government Working Group, 2001).

The need for different standards is also driven by the many different views of what is important about the information that the metadata describes. As Gilliland explains (Gilliland, 2016), there are different notions of provenance, different institutional cultures, and divergent cultural approaches. This leaves "many professionals, and the communities they represent, feeling that their practices and needs have been shoehorned into structures that were developed by another community with quite different epistemologies, practices, and users." Libraries and archives, for example, sometimes use different standards to describe the exact same kind of information (Elings & Waibel, 2007).

As useful and even necessary as complex and specialized standards can be, their very complexity and specialization can be problematic. Very complex metadata standards can be difficult (and, therefore, expensive) to implement accurately, and that can lead to the standards being underutilized (Ulrich et al., 2022).

Indeed, one reason cited for the failure of the *Government Information Locator Service* (GILS) standard was its complexity (Doyle, 2003). Introduced in 1994 and instantiated into law as part of the *Paperwork Reduction Act of 1995* (44 USC 3511) as a way to describe publicly available federal information resources (GPO, 2007a), GILS was pretty much dead by 2008 (GPO, 2010).

In spite of the challenges, some standards do become successful and widely used. To a very real extent, success of a standard is a chicken-and-egg problem: a standard becomes widely used if it is successful, but in order to become successful, a standard must be widely used. Success leading to more use is a classic example of the "network effect" in which lots of people use something because lots of people use it (Weinberger & Searls, 2015; US President, 2023).

The contrasting costs and benefits of simple and complex metadata standards present a challenge. Use of simple, generic metadata standards can lead to their wide adoption. The Dublin Core standard is an example of this (Westin, 2024; Search.gov, 2020; Fretwell, 2024). But simple, generic metadata cannot adequately describe complex types of information, nor does it necessarily meet the use-needs of specific communities. An understanding of how traditional metadata for the paper era does and does not work in the digital age will help preservationists meet this challenge.

Paper-Era Metadata

In the paper era, archives and libraries used metadata to describe physical objects that contained information. For libraries, the objects were mostly books and journals; for archives, the objects were typically groups of documents (fonds, collections, series, record groups) linked by provenance and stored in archival Hollinger boxes. This was the "container" approach: whether the container was the physical volume of a single book, or many volumes of a single serial title, or a single collection stored in many boxes, a single record described the contents of the container. The granularity of the description of the information in the container varied greatly for different kinds of containers.

It made sense to combine these overlapping functions into a single record for each unique item in a library (or collection in an archive) because information was distributed, stored, managed, and used in the same physical containers described by the single record. Users looked for and retrieved the same objects that the library and archive managed. The physicality of the information packaging drove the design of metadata.

In that environment, libraries and archives used the same metadata records for two very different functions: to manage their collections and to help their users find and locate the information they wanted. Although these two functions and the meta-information they require overlap, the functions are very different.

In his textbook on metadata, Jeffrey Pomerantz describes the paper-era metadata model as the "One-to-One Principle: one resource, one record" (Pomerantz, 2015). This principle also had a side effect where "related but conceptually different entities, for example a painting and a digital image of the painting, are described by separate metadata records" (Woodley, 2005). The result of this model was that each metadata record was inextricably linked to the information object it described but was isolated from related records and that, in turn, isolated the information itself from related information.

The single-record, one-to-one metadata model is inadequate in the digital age and can even be a barrier to progress. It imposes constraints on discovery that, in the digital age, are unnecessary and limiting. This old metadata model needs to be supplemented by new models to overcome those limitations.

This can be understood by looking at four specific limitations of the one-to-one model

First, it is a "choke point" that literally limits what metadata can be used for discovery. Using only a single-record model forces all metadata into the elements, rules, and encodings of a single metadata standard.

Second, the single-record model is designed to describe, index, and retrieve information containers rather than the information in the containers. In the digital age, the inherent characteristics of digital information often makes it possible and even preferable to deliver to users specific units of information rather than their published containers.

Third, the single-record model of metadata has practical limitations that make it hard to scale to the volume of digital information being produced by government agencies. GPO explicitly acknowledged this fact in the early days of its experimentation with web harvesting, estimating that it would take four years to catalog the harvest from a single agency (Federal Depository Library Council, 2007). GPO coped with this scaling problem by switching from using a single record to describe a single document to using a single record to describe whole collections of documents. GPO's general policy for its FDLP Web Archive project (GPO, 2021a) is to create a single catalog record for an entire website of files and documents and web pages (FDLP, 2023c). The result is that, as of fiscal year 2023, GPO used just 269 records in the *Catalog of Government Publications* to describe 555 million URLs and over 52 terabytes of data (GPO, Library Services & Content Management, 2024). This drastically minimizes the number of access points and diminishes discoverability at a time when user expectations are for more and better accessibility.

Finally, limiting metadata to descriptions of individual documents (much less to whole series or whole websites) is simply no longer adequate. Ted Nelson described our information environment as everything being "deeply intertwingled" (Nelson, 1975). Sven Birkerts said that "the book" is now the network and the network is now knowledge (Birkerts, 1998). The digital age is no longer about books that are isolated from each other; the digital age is about information that is connected in myriad ways. In the language of the semantic web, the world wide web is no longer a web of documents; it is a web of data (US Library of Congress, 2012).

In this world, using limited resources to support only the old metadata model takes resources away from supporting innovative forms of metadata that can empower libraries and archives to build digital user-services that were simply not possible in the paper era.

Digital-Age Metadata

Digital archives and libraries can use new models of metadata that will support the traditional functions of identifying, managing, and supporting discovery, but that will also enable better discovery and new functionality. This will require metadata that is detailed but also flexible and reusable in creative ways.

Although the metadata for any given function often overlaps with the metadata needed for (or usable for) other functions, it is nevertheless reasonable to conclude that no single metadata standard can serve every function for every library or archive or Designated Community. At best, a single metadata model would have to be a compromise among competing functions and interests and values.

A successful DPI will need a metadata model that can be efficient, widely adopted, easily implemented, and flexible. But most importantly, it must be designed in such a way as to make it easy to gather and generate and create metadata at an enormous scale.

Evolution of Metadata

A lot has changed in the world of metadata over the lasts 20 years. New ideas and assumptions about how metadata are created and used have led to new standards and new ways of thinking about metadata.

Gordon Dunsire and colleagues have suggested, for example, that the old monolithic, "top-down" approach to metadata "is not going to allow us to take advantage of new technologies or new ways of thinking about and building metadata" (Dunsire et al., 2017). They suggest that a "bottom-up" approach would change the focus from "controlling the data" to "controlling the semantics" of expressing the data. Doing so would allow *users* to re-use metadata. It would allow metadata "to function for any purpose imaginable. What those new functions might be, and how well library metadata serves them, should trouble libraries no more than how a patron uses the information from their collections." The "true worth" of such metadata would be its consistency and completeness.

The creation of the linked-data model is one of the most significant technological developments in cataloging (Dobreski, 2021). One of its main innovations is the change of the encoding of bibliographic descriptive metadata from the outdated MARC model to the Resource Description Framework model. RDF enables the metadata to use "statements" rather than "records." Statements can be used iteratively, with an object of one statement becoming the subject of another statement. RDF makes it possible to repurpose individual elements of metadata, making it possible to create composite records made up of metadata from different sources. The linked data model is critical to the idea of the semantic web (Berners-Lee, 2006).

The 2008 Library of Congress report, *On the Record* (Library of Congress Working Group on the Future of Bibliographic Control, 2008), suggested that a metadata infrastructure should allow the "display and indexing of data elements to vary according to the particular needs of the communities concerned." The report also recognized that library metadata should be integrated into the web rather than isolated in catalog databases. In this future, the web itself will be the platform for the delivery of library metadata. This means that library and archival metadata will be available directly on the web rather than isolated in library databases accessible on the web. Karen Coyle, one of the report's consultants, has summarized this idea succinctly: data must be *of the web* and not just *on the web* (Guerrini, 2023).

Rethinking Metadata

The changes described above can guide how preservationists might think about metadata for a DPI. We list here some ideas.

The old model was simple but not very flexible. A new model would be more flexible by allowing any element to be recorded accurately and completely and used selectively based on whatever function a library or archive wished to provide. Libraries could then select the elements for a particular function and mix and match them in creative ways. Such a model would not prevent libraries and archives from using metadata the old way, but it would enable libraries and archives to develop new innovative services more easily than they could under the old model. In doing so, it might just enable new ways of implementing the old model as well.

Classifying Elements

There are thousands of properties of information that metadata can describe, but they are not all alike. Grouping elements into broad categories that share common characteristics provides an opportunity for thinking about metadata from the ground up instead of from the top down.

There are at least three distinct *kinds* of metadata elements.

- Elements (such as PID) that have unique values for each individual item of information.
- Elements (such as file-type and author) that have values that more than one (often many) items of information share.
- Elements that indicate relationships between items of information. For example, different editions or formats or manifestations of a work are all related to each other (Tillett, 2004). And an item (such as a statistical table or an image) within another item (such as a journal article) is related to its parent item (the journal article) and to its sibling items (other tables and images and text in the article). Other examples: an item of testimony in a hearing, an announcement in an issue of the *Federal Register*, a section of a Statute, an amicus brief in a court case. And, of course, most web pages point to and are pointed to by other web pages.

Given that there are different categories of metadata elements, it should be more efficient if the rules for recording metadata values are designed to conform to the unique characteristics of each type. The old metadata model relies on one record for each book (or other separately published information object). A new metadata model could rely on one record for each element. A book-record has to focus on describing the published information object; an element-record could focus on the

373

unique nature of each individual element while still connecting the element to each appropriate information item.

Building Element-Records

A model that allowed the building of element-records could work by relying on Persistent Identifiers. Providing each information object with a unique PID would allow the PID to serve as a Primary Key that links different elements together.

This would enable preservationists to construct different kinds of metadata files (or "element records") such as:

- **Simple lists**. For example, a "PDF list" could consist of a list of all the PIDs of all described information objects available as PDF files. A "collection list" could consist of all the PIDs for a library's Defined Collection. (This would make it easy for libraries to construct and manage more than one Defined Collection, each for a different Community.) With a central warehouse of metadata and publicly available metadata tools and APIs (Application Programming Interfaces) made available to the public, users could create their own lists of their own selections of information. Bibliographies could be dynamically constructed from simple lists of PIDs. Lists could be fixed and permanent, or fixed and growing with new items added as needed, or dynamic and based on selected and changing criteria.

- **Key-Value lists**. For example, a "title list" could consist of a list of each title and its PID. (Functionally, the keys and values can be used interchangeably. Thus, in this example, the title can be used to retrieve the PID or the PID can be used to retrieve the title.)

- **Arrays**. For example, an "author array" file could consist of a list of authors with one array for each. Each author's name (or ID) would contain an array (i.e., a list) of all the author's information items (Author: list of PIDs). A "location array" could keep track of information items that moved around during their lifespan on the web and could consist of an array of old URLs associated with an information item (PID: list of old URLs). Such a location array would make it possible to design a service for finding a preserved copy of an item with a broken link. Complex or "associative" arrays could include a date for each item allowing, the easy retrieval of "mementos" (which are different versions of the same item) (Van de Sompel et al., 2009). A "manifestations array" could consist of an array of the PIDs of each "manifestation" of a "work" (work: PIDs) (IFLA Study Group on the Functional Requirements for Bibliographic Records, 2009). Holdings data could be stored in an array for each serial title.

- **Triples**. The long-promised "semantic web" applications could emerge from centrally stored and shared metadata that expressed relationships between items. "Triples" are statements that connect three pieces of information: a "subject" connected to an "object" with a "predicate" (Coyle, 2010; Hyland & Wood, 2011). Triples could be used to preserve the original links between items on the web (e.g., "document A" "was linked to by" "whitehouse.gov"). This could help classify content and be used to weight search results.

The old model relied on complex commercial software that was designed to read and understand a single complex data format. This allowed librarians and archivists to create, add, modify, and use metadata by using the commercial software.

A new model could store elements and their values using common data structures (like lists, arrays, and triples) serialized into common formats (such as YAML, JSON, and XML). This model would allow librarians and archivists to use common programming languages (which natively support these common data structures and formats) to select, combine, mix, and re-mix metadata from the common store of metadata into digital services using the database and web application software of their choice. Libraries and archives would no longer have to rely only on expensive commercial software with its fixed range of functions. Instead, they could use the database or web application software most appropriate for each digital service. Digital services would not be tied to relational databases or the database structure used by OPAC and ILS (Integrated Library System) vendors. Each digital service could use the type of database (e.g., NoSQL, NewSQL, columnar, graph, object-oriented, vector) best suited to a particular service.

Even the old-model book-records that describe a single item thoroughly and completely could be assembled from necessary elements or displayed or delivered dynamically from requested elements.

Bibliography

Laws, Regulations, Court Decisions

PUBLIC LAWS

An Act To provide for the disposition of certain records of the United States Government, Pub. L. No. 295 (53 Stat 1219) [CHAPTER 481] (1939). https://heinonline.org/HOL/P?h=hein.statute/sal053&i=1505

Depository Library Act of 1962, Pub. L. No. 87-579, 76 Stat. 352. https://uscode.house.gov/statutes/pl/87/579.pdf

E-Government Act of 2002, Pub. L. No. 107-347 (116 Stat. 2899). https://www.govinfo.gov/content/pkg/PLAW-107publ347/pdf/PLAW-107publ347.pdf.

Electronic Freedom of Information Act Amendments, Pub. L. No. 104-231, 110 Stat. 3048 (1996). https://uscode.house.gov/statutes/pl/104/231.pdf

Federal Records Act, Pub. L. No. 81-754 [§ 6], 64 Stat. 578, (1950).

Freedom of Information Act, Pub. L. No. 89-487, 80 Stat. 250, (1966) https://www.govinfo.gov/content/pkg/STATUTE-80/pdf/STATUTE-80-Pg250.pdf https://www.law.cornell.edu/uscode/text/5/552

Government Printing Office Electronic Information Access Enhancement Act of 1993, Pub. L. No. 103-40, 107 Stat 112. http://www.gpo.gov/fdsys/pkg/STATUTE-107/pdf/STATUTE-107-Pg112.pdf

The Judiciary Appropriations Act, 1991, Pub. L. No. 101-515, 104 Stat. 2129 (1990). https://www.congress.gov/101/statute/STATUTE-104/STATUTE-104-Pg2101.pdf

Public Printing and Documents, Pub. L. No. 90-620, 82 Stat 1238, (1968) https://www.govinfo.gov/content/pkg/STATUTE-82/pdf/STATUTE-82-Pg1238.pdf

SUPREME COURT

Immigration and Naturalization Service v. Chadha, 462 US 919 (1983) https://www.loc.gov/item/usrep462919/

CODE OF FEDERAL REGULATIONS

Creation And Maintenance Of Federal Records. 36 C.F.R. § 1222 (2024). https://www.ecfr.gov/current/title-36/chapter-XII/subchapter-B/part-1222

What types of documentary materials are Federal records?. 36 C.F.R. § 1222.12 (2024). https://www.ecfr.gov/current/title-36/chapter-XII/subchapter-B/part-1222/subpart-A/section-1222.12

UNITED STATES CODE

2 USC 145-146. 2 US Code Chapter 5 - LIBRARY OF CONGRESS. https://www.law.cornell.edu/uscode/text/2/chapter-5

5 USC 552. Public information; agency rules, opinions, orders, records, and proceedings. https://www.law.cornell.edu/uscode/text/5/552

44 USC 7. CONGRESSIONAL PRINTING AND BINDING. https://www.law.cornell.edu/uscode/text/44/chapter-7

44 USC 17 DISTRIBUTION AND SALE OF PUBLIC DOCUMENTS. https://www.law.cornell.edu/uscode/text/44/chapter-17

44 USC 1710 Index of documents: number and distribution. https://www.law.cornell.edu/uscode/text/44/1710

44 USC 1711 Catalog of Government publications. https://www.law.cornell.edu/uscode/text/44/1711

44 USC 1714 Publications for use of National Archives and Records Administration. https://www.law.cornell.edu/uscode/text/44/1714

44 USC 1718 Distribution of Government publications to the Library of Congress. https://www.law.cornell.edu/uscode/text/44/1718

44 USC 19 DEPOSITORY LIBRARY PROGRAM. https://www.law.cornell.edu/uscode/text/44/chapter-19

44 USC 1901 Definition of Government publication. https://www.law.cornell.edu/uscode/text/44/1901

44 USC 1902 Availability of Government publications through Superintendent of Documents; lists of publications not ordered from Government Publishing Office. https://www.law.cornell.edu/uscode/text/44/1902

44 USC 1911 Free use of Government publications in depositories; disposal of unwanted publications. https://www.law.cornell.edu/uscode/text/44/1911

44 USC 1912 Regional depositories; designation; functions; disposal of publications. https://www.law.cornell.edu/uscode/text/44/1912

44 USC 2108 Responsibility for custody, use, and withdrawal of records. https://www.law.cornell.edu/uscode/text/44/2108

44 USC 2118 Records of Congress. https://www.law.cornell.edu/uscode/text/44/2118

44 USC 2201 Ownership of Presidential records. https://www.law.cornell.edu/uscode/text/44/2202

44 USC 27 ADVISORY COMMITTEE ON THE RECORDS OF CONGRESS. https://www.law.cornell.edu/uscode/text/44/chapter-27

44 USC 2901 RECORDS MANAGEMENT BY THE ARCHIVIST OF THE UNITED STATES AND BY THE ADMINISTRATOR OF GENERAL SERVICES : Definitions. https://www.law.cornell.edu/uscode/text/44/2901

44 USC 3301 DISPOSAL OF RECORDS : Definition of records. https://www.law.cornell.edu/uscode/text/44/3301

44 USC 3303 Lists and schedules of records to be submitted to the Archivist by head of each Government agency. https://www.law.cornell.edu/uscode/text/44/3303

44 USC 3502 FEDERAL INFORMATION POLICY : Definitions. https://
www.law.cornell.edu/uscode/text/44/3502

44 USC 3511 Data inventory and Federal data catalogue. https://
www.law.cornell.edu/uscode/text/44/3511

44 USC 41 ACCESS TO FEDERAL ELECTRONIC INFORMATION. https://
www.law.cornell.edu/uscode/text/44/chapter-41

44 USC 4101 Electronic directory; online access to publications;
electronic storage facility. https://www.law.cornell.edu/uscode/text/
44/4101

44 USC 4102 ACCESS TO FEDERAL ELECTRONIC INFORMATION :
Fees. https://www.law.cornell.edu/uscode/text/44/4102

Books, Articles, Reports

Abrams, S., Kunze, J., & Loy, D. (2010). An Emergent Micro-Services
Approach to Digital Curation Infrastructure. *The International Journal
of Digital Curation*, *1*(5), 173–186. http://www.ijdc.net/article/view/
154

Acharya, A. (2009). Finding the laws that govern us. *Official Google
Blog*. https://googleblog.blogspot.com/2009/11/finding-laws-that-
govern-us.html

Administrative Office of the U.S. Courts. (2000). Electronic Public Access
at 10. *The Third Branch*. https://web.archive.org/web/
20010306002030/http://www.uscourts.gov/ttb/sept00ttb/epa.html

Administrative Office of the U.S. Courts. (2005). NEW! Free Written
Opinions. *PACER Service Center Announcements*. https://perma.cc/
295D-DPHH

Administrative Office of the U.S. Courts. (2011). New Pilot Project Will
Enhance Public Access to Federal Court Opinions. *News Release*.
https://web.archive.org/web/20110505031658/https://
www.uscourts.gov/News/NewsView/11-05-04/

New_Pilot_Project_Will_Enhance_Public_Access_to_Federal_Court_Opinions.aspx

Administrative Office of the U.S. Courts. (2013a). Access to Court Opinions Expands. *News Release*. https://www.uscourts.gov/news/2013/01/31/access-court-opinions-expands

Administrative Office of the U.S. Courts. (2013b). Reappraisal of Records Saving Millions for Judiciary. *Judiciary News*. https://www.uscourts.gov/news/2013/08/20/reappraisal-records-saving-millions-judiciary

Administrative Office of the U.S. Courts. (2016). *The Federal Court System In The United States* (4th ed.). https://www.uscourts.gov/sites/default/files/federalcourtssystemintheus.pdf

Administrative Office of the U.S. Courts. (2022a). *Judicial Administration*. https://www.uscourts.gov/about-federal-courts/judicial-administration

Administrative Office of the U.S. Courts. (2022b). PACER Case Locator (PCL) User Manual. https://pacer.uscourts.gov/sites/default/files/files/PACER%20Case%20Locator%20User%20Manual_0.pdf

Administrative Office of the U.S. Courts. (2022c). PACER User Manual for CM/ECF Courts. https://pacer.uscourts.gov/sites/default/files/files/PACER-User-Manual.pdf

Administrative Office of the U.S. Courts. (2025). *PACER Pricing: How fees work*. https://pacer.uscourts.gov/pacer-pricing-how-fees-work

Ahmad, R., Rafiq, M., & Arif, M. (2023). Global trends in digital preservation: Outsourcing versus in-house practices. *Journal of Librarianship and Information Science*. https://doi.org/10.1177/09610006231173461

Albaum, S. (2002). The National Reporter System Celebrates Historic Anniversary. *LLRX*. https://web.archive.org/web/20100111064033/http://www.llrx.com/features/nationalreporter.htm

Alliance for Permanent Access to the Records of Science Network. (2012). Report On Peer Review Of Digital Repositories. http://www.alliancepermanentaccess.org/wp-content/uploads/downloads/2012/04/APARSEN-REP-D33_1B-01-1_0.pdf

American Bar Association. (2020). Free Access to Court Records. *Washington Letter*. https://www.americanbar.org/advocacy/ governmental_legislative_work/publications/washingtonletter/ dec-2020-wl/pacer-wl1220/

American Library Association. (2025). *Key Principles of Government Information*. https://www.ala.org/advocacy/govinfo/keyprinciples

Ammerman, J. (2020). Discovery, Access, and Use of Information in a "Digital Ecosystem". In S. L. Mizruchi (Ed.), *Libraries and archives in the digital age* (pp. 43-50). Palgrave Macmillan. https://doi.org/ 10.1007/978-3-030-33373-7

Anderson, M. (2011a). E is for ecology. *The Signal*. https://blogs.loc.gov/ thesignal/2011/10/e-is-for-ecology/

Anderson, R. (2011b). As Book Warehouses Vanish, Is It Time for Librarians to Stop Running Libraries? *The Scholarly Kitchen*. https:// scholarlykitchen.sspnet.org/2011/05/23/as-book-warehouses-vanish- is-it-time-for-librarians-to-stop-running-libraries/

Anderson, R. (2011c). What Patron-Driven Acquisition (PDA) Does and Doesn't Mean: An FAQ. *The Scholarly Kitchen*. https:// scholarlykitchen.sspnet.org/2011/05/31/what-patron-driven- acquisition-pda-does-and-doesnt-mean-an-faq/

Anderson, R. (2013). *Can't buy us love: The declining importance of library books and the rising importance of special collections*. Ithaka S+R Issue Briefs. https://doi.org/10.18665/sr.24613

Arrigo, P. A. (2004). The reinvention of the FDLP: A paradigm shift from product provider to service provider. *Journal of Government Information*, *30*(5-6), 684–709. https://doi.org/10.1016/ j.jgi.2004.11.003

Association of Research Libraries. (2009). Strategic Directions for the Federal Depository Library Program. *White Paper*. https://www.arl.org/ wp-content/uploads/2009/04/fdlp-strategic-directions-april09.pdf

Association of Southeastern Research Libraries. (2012). *Best Practices Documentation: Becoming a Center of Excellence for ASERL's*

Collaborative Federal Depository Program. https://www.aserl.org/wp-content/uploads/2012/06/BestPractices-Update_2012_05.pdf

Association of Southeastern Research Libraries. (2018). Southeast Region Guidelines For Management And Disposition Of Federal Depository Library Collections, Revised July 2018. https://www.aserl.org/wp-content/uploads/2018/10/ASERL_FDLP_GUIDELINES_Revised_Final_2018_09.pdf

Baldwin, G. (1997). The Electronic Transition - A Progress Report. *Administrative Notes, 18*(5), 11–20. https://www.govinfo.gov/content/pkg/GOVPUB-GP3-49b8d6fa12ef7aa08141bf059deafa16/pdf/GOVPUB-GP3-49b8d6fa12ef7aa08141bf059deafa16.pdf

Baldwin, G. (2000). Library Programs Service Update. *Administrative Notes, 21*(7), 7–12. https://www.govinfo.gov/content/pkg/GOVPUB-GP3-86a923c932696e39e1c1d36232ead077/pdf/GOVPUB-GP3-86a923c932696e39e1c1d36232ead077.pdf

Baldwin, G. (2003). Fugitive Documents – On the Loose or On the Run. *Administrative Notes, 24*(10), 4–8. https://www.govinfo.gov/content/pkg/GOVPUB-GP3-849c445ba7debf10b1c116284b6aa82b/pdf/GOVPUB-GP3-849c445ba7debf10b1c116284b6aa82b.pdf

Balter, B. J. (2011). *Analysis of Federal Executive .Govs*. https://web.archive.org/web/20111101142750/http://ben.balter.com/2011/09/07/analysis-of-federal-executive-domains

Banerjee, K., & Forero, D. (2017). Leveraging the DOI Infrastructure to Simplify Digital Preservation and Repository Management. *Code{4}Lib*, 38). http://journal.code4lib.org/articles/12870

Barnum, G. (2011). *Public Documents Our Specialty: GPO's Public Documents Library 1895-1971 Part 2: Changing Times*. https://web.archive.org/web/20151021085514/https://www.fdlp.gov/all-newsletters/featured-articles/1142-public-documents-library-part2

Barnum, G. D., & Kerchoff, S. P. (2000). The federal depository library program electronic collection: Preserving a tradition of access to United States government information. *New Review of Academic Librarianship, 6*(1), 247–255. https://doi.org/10.1080/13614530009516814

Bibliography

Bartram, A. (2001). *Five hundred years of book design.* Yale University Press. https://search.worldcat.org/en/title/48260555

Bates, M. J. (2018). Information. In J. D. McDonald & M. Levine-Clark (Eds.), *Encyclopedia of Library and Information Science* (4th ed.). CRC Press. https://doi.org/10.1081/E-ELIS4

Bateson, G. (1972). *Steps to an ecology of mind; collected essays in anthropology, psychiatry, evolution, and epistemology.* Chandler Pub. Co. https://search.worldcat.org/en/title/258014

Baucom, E. (2019). A Brief History of Digital Preservation. *Mansfield Library Faculty Publications, 31.* https://scholarworks.umt.edu/ml_pubs/31/

Becker, C. (2018). Metaphors we work by: reframing digital objects, significant properties, and the design of digital preservation systems. *Archivaria, 85,* 6–37. https://archivaria.ca/index.php/archivaria/article/download/13628/15017/

Behnke, J., Mitchell, A., & Ramapriyan, H. (2019). NASA's Earth observing data and information system--near-term challenges. *Data Science Journal, 18,* 40–40. https://doi.org/10.5334/dsj-2019-040

Bennett, S., & Council on Library and Information Resources. (2005). *Library as place : rethinking roles, rethinking space.* https://www.clir.org/wp-content/uploads/sites/6/pub129.pdf

Bergman, A., & Rinberg, T. (2018). *In its first year, the Trump administration has reduced public information online.* https://sunlightfoundation.com/2018/01/04/in-its-first-year-the-trump-administration-has-reduced-public-information-online/

Berners-Lee, T. (1996). The Name Myth -- Axioms of Web architecture. *w3.org.* https://www.w3.org/DesignIssues/NameMyth.html

Berners-Lee, T. (1998). Cool URIs don't change. *w3.org.* http://www.w3.org/Provider/Style/URI.html.en

Berners-Lee, T. (2006). *Linked Data.* https://www.w3.org/DesignIssues/LinkedData.html

Bethany, L. (2018). Digital Government Information: The Challenges of Collaborative Preservation. *The Journal of Academic Librarianship, 44*(5), 674–676. https://doi.org/10.1016/j.acalib.2018.07.005

Birkerts, S. (1998). The book is the network, the network is knowledge. *Atlantic Unbound.* http://www.theatlantic.com/past/unbound/digicult/dc980910.htm

Biscardi, F. (1993). The Historical Development of the Law Concerning Judicial Report Publication. *Law Library Journal, 85*(3). https://heinonline.org/HOL/LandingPage?handle=hein.journals/llj85&div=43&id=&page=

Bleakly, D. R. (2002). *Long-Term Spatial Data Preservation and Archiving: What are the Issues?* (Vol. SAND 2002-0107). Sandia National Laboratories. https://www.osti.gov/servlets/purl/793225

Borgman, C. L. (2003). The Invisible Library: Paradox of the Global Information Infrastructure. *Library Trends, 51*(4), 652–674. http://hdl.handle.net/2142/8487

Borgman, C. L. (2007a). Building an Infrastructure for Information. In *Scholarship in the Digital Age: Information, Infrastructure, and the Internet* (pp. 149-177). MIT Press. https://ieeexplore.ieee.org/abstract/document/6279958

Borgman, C. L. (2007b). Scholarship in the Digital Age: Information, Infrastructure, and the Internet. https://doi.org/10.7551/mitpress/7434.001.0001

Borodin, Y., Bigham, J. P., Dausch, G., & Ramakrishnan, I. V. (2010). More than meets the eye: a survey of screen-reader browsing strategies. In *Proceedings of the 2010 International Cross Disciplinary Conference on Web Accessibility (W4A)* (pp. 1-10). https://cs.rochester.edu/hci/pubs/pdfs/browsing-strategies-w4a10.pdf

Boté-Vericad, J.-J., & Healy, S. (2022). Academic Libraries And Research Data Management: A Systematic Review. *Vjesnik bibliotekara Hrvatske, 65*(3), 171–171. https://www.academia.edu/download/96612414/Bote_Healy_Research_Data_Management_S_Review.pdf

Bower, D., Gao, F., & Walls, D. (2012). Web Archiving for the FDLP. *Depository Library Conference.* https://web.archive.org/web/20160916215327if_/https://www.fdlp.gov/file-repository/outreach/events/depository-library-council-dlc-meetings/2012-meeting-proceedings/2229-web-archiving-and-how-it-fits-into-the-fdlp/file

Bower, D., & Walls, D. (2014). *Capturing and Preserving the Federal Web for Permanent Public Access.* In. https://web.archive.org/web/20160917042510/https://www.fdlp.gov/file-repository/outreach/events/depository-library-council-dlc-meetings/2014-meeting-proceedings/depository-library-council-meeting-and-federal-depository-library-conference-april-30-may-2-2014/2438-bringing-order-to-chaos-capturing-the-federal-web-for-permanent-public-access/file

Bowker, G. C., & Star, S. L. (1999). *Sorting Things Out : Classification and Its Consequences.* MIT Press. https://hcommons.org/app/uploads/sites/1001532/2020/03/Bowker-1999-Sorting-Things-Out-Classification-and-Its-Consequences.pdf

Braddock, J. (2021). *FDLP Web Archive Resources.* https://web.archive.org/web/20210701122356/https://www.fdlp.gov/cataloging-guidelines/bibliographic-cataloging/continuing-resources/integrating-resources/fdlp-web-archive-resources

Brown, T. E. (2006). Toward the Appraisal of Web Records. *Archival Outlook: Newsletter of the Society of American Archivists, 6,*25. http://files.archivists.org/periodicals/Archival-Outlook/Back-Issues/2006-4-AO.pdf

Buckland, M. K. (1991). Information as thing. *Journal of the American Society for information science, 42*(5), 351–360. https://escholarship.org/content/qt4x2561mb/qt4x2561mb_noSplash_aae8ca6ae9c862efc2341bb6ae3d855f.pdf

Buckland, M. K. (1997). What is a "document". *Journal Of The American Society For Information Science, 48*(9), 804–809. https://www.marilia.unesp.br/Home/Instituicao/Docentes/EdbertoFerneda/pd-what-is-a-document.pdf

Buckley, F. J. (2000a). Update on Superintendent of Documents Programs. *Administrative Notes, 21*(7), 1–6. https://www.govinfo.gov/content/pkg/GOVPUB-GP3-86a923c932696e39e1c1d36232ead077/pdf/GOVPUB-GP3-86a923c932696e39e1c1d36232ead077.pdf

Buckley, F. J. (2000b). Letter to Depository Library Directors (8/25/2000). *govdoc-l.* https://lists.psu.edu/cgi-bin/wa?A2=GOVDOC-L;602528b.00&S=a

Burger, J., Gherman, P. M., & Wilson, F. (2005). *ASERL's Virtual Storage/Preservation Concept.* https://web.archive.org/web/20140424202330/http://www.ala.org/acrl/sites/ala.org.acrl/files/content/conferences/pdf/burger-etal05.pdf

Burke, A. (2023). Reader Mode: Sweeping Away Barriers to Reading. *LANGUAGE TEACHER, 47*, 51. https://jalt-publications.org/sites/default/files/pdf-article/47.6tlt-wired_0.pdf

Burton, M. (2015). *Blogs as Infrastructure for Scholarly Communication.* https://deepblue.lib.umich.edu/handle/2027.42/111592.

Bustillos, M. (2022). Archiving official documents as an act of radical journalism. *Columbia Journalism Review.* https://www.cjr.org/business_of_news/archiving-official-documents-as-an-act-of-radical-journalism.php

Butler, J. (1987). Collection building vs. document delivery: an evaluation of methods to provide NTIS documents in an academic engineering library. *Report No: ED 286 510.* https://eric.ed.gov/?id=ED286510

Calma, J. (2022). The EPA plans to sunset its online archive. *The Verge.* https://www.theverge.com/2022/3/24/22993628/epa-online-archive-sunset-digital-records

Case, D. O. (2002). *Looking for information : a survey of research on information seeking, needs, and behavior.* Academic Press.

Center for Research Libraries. (2010). Report on Portico Audit Findings. https://live-crl-www.pantheonsite.io/sites/default/files/reports/CRL%20Report%20on%20Portico%20Audit%202010.pdf

Center for Research Libraries. (2025). Technical Report Archive & Image Library. https://www.crl.edu/programs/trail

Chandler, Y. J. (2000). Legal Information on the Internet. *Journal of Library Administration, 30*(1-2), 157–207. https://doi.org/10.1300/J111v30n01_03

Chesapeake Digital Preservation Group. (2013). "Link Rot" and Legal Resources on the Web. https://web.archive.org/web/20170702052408/https://library.law.uiowa.edu/sites/library.law.uiowa.edu/files/Link%20Rot%20and%20Legal%20Resources%20on%20the%20Web.pdf

Chin, M. (2021). File not found. *The Verge.* https://www.theverge.com/22684730/students-file-folder-directory-structure-education-gen-z

Claburn, T. (2009). Government Keeping Its .Gov Domain Names Secret. *Information Week.* https://web.archive.org/web/20090306121859/http://www.informationweek.com/news/showArticle.jhtml?articleID=215600330

Clerk of the House [US]. (2024). [Report on adopting standardized formats for legislative documents]. *Quarterly Report, 12.* https://usgpo.github.io/innovation/resources/reports/Clerk-QR12-Standardized-Formats.pdf

Commission on Preservation and Access. (1996). The Commission on Preservation and Access, The Council on Library Resources: Annual Report, July 1, 1995-June 30, 1996. https://www.clir.org/wp-content/uploads/sites/6/2020/10/1995-1996-Annual-Report.pdf

Committee on the Records of Government, Social Science Research Council, American Council of Learned Societies, & Council on Library Resources. (1985). *Committee on the records of government : report.* [American Council of Learned Societies] : [Council on Library Resources]: [Social Science Research Council]. https://eric.ed.gov/?id=ED269018

Congressional Research Service. (2017). Government Printing, Publications, and Digital Information Management: Issues and

Challenges. *CRS Report, R45014*. https://www.everycrsreport.com/reports/R45014.html

Congressional Research Service. (2021). Federal Workforce Statistics Sources: OPM and OMB. *CRS Report, R43590*. https://sgp.fas.org/crs/misc/R43590.pdf

Consultative Committee for Space Data Systems (CCSDS). (1999). Reference Model for an Open Archival Information System (OAIS). *CCSDS 650.0-R-1*. https://www.tc.faa.gov/its/worldpac/standards/oais%20CCSDS-650.0-R-2.pdf

Consumer Financial Protection Bureau. (2019). Request for Records Disposition Authority [Approved]. *DM-0587-2018-0001*. https://www.archives.gov/files/records-mgmt/rcs/schedules/independent-agencies/rg-0587/daa-0587-2018-0001_sf115.pdf

Conway, P. (1996). *Preservation in the Digital World*. Report. http://www.clir.org/pubs/reports/conway2/index.html

Cornell University Library. (2003). Digital Preservation Strategies. http://www.dpworkshop.org/dpm-eng/terminology/strategies.html

Corrado, E., M. (2019). Repositories, Trust, and the CoreTrustSeal. *Technical Services Quarterly, 36*(1), 61–72. https://doi.org/10.1080/07317131.2018.1532055

Cox, R. J. (2013). Archival Futures: The Future of Archives. *Collections, 9*(4), 331–352. https://doi.org/10.1177/155019061300900402

Coyle, K. (2010). Changing the Nature of Library Data. *Library Technology Reports, 46*(1), 14–29. https://journals.ala.org/index.php/ltr/article/view/4629/5475

Cruse, P., Eckman, C., Kunze, J., & Christenson, H. (2003). Web-Based Government Information: Evaluating Solutions for Capture, Curation, and Preservation. http://www.cdlib.org/services/uc3/docs/Web-based_archiving_mellon_Final.pdf

Cybersecurity and Infrastructure Security Agency. (2022). *dotgov-data, current-federal.csv*. https://github.com/cisagov/dotgov-data/blob/main/current-federal.csv

Data Refuge. (2016). DataRefuge. https://perma.cc/7MBM-7HJS

Data Rescue Project. (2025a). *The Data Rescue Tracker*. https:// www.datarescueproject.org/data-rescue-tracker/

Data Rescue Project. (2025b). https://www.datarescueproject.org/

Davis, M. E. (1996). Cornell named site for federal labor reports. *ACRL College & Research Libraries News, 57*(3). https://crln.acrl.org/ index.php/crlnews/article/view/19761/23428

Davis, R. (2009). Update [for the Federal Documents Task Force Meeting, ALA 2009 Midwinter]. *Administrative Notes, 30*(1), 7–18. https:// www.govinfo.gov/app/details/GOVPUB- GP3-884535854f266455cd1a1df7e7fa87b1

Davis, R. (2008). What Will the Future of the FDLP Look Like? *Administrative Notes, 29*(07-08), 4–9. https://www.govinfo.gov/ content/pkg/GOVPUB-GP3-97e8dada618d6ffdbb435b5851522c07/ pdf/GOVPUB-GP3-97e8dada618d6ffdbb435b5851522c07.pdf

DCMI-Government Working Group. (2001). *DC-GOV Application Profile*. https://www.dublincore.org/groups/government/ profile-200202/

Dempsey, L. (2017). Library Collections in the Life of the User: Two Directions. *Liber Quarterly, 26*(4), 338–359. https://liberquarterly.eu/ article/download/10870/11778

Dempsey, L., Malpas, C., & Lavoie, B. (2014). Collection Directions: Some Reflections on the Future of Library Collections and Collecting. *portal: Libraries and the Academy, 14*(3), 393–423. https://doi.org/ 10.25333/tw3v-xb93

Dennis, B. (2016). Scientists are frantically copying U.S. climate data, fearing it might vanish under Trump. *Washington Post*. https:// www.washingtonpost.com/news/energy-environment/wp/2016/12/13/ scientists-are-frantically-copying-u-s-climate-data-fearing-it-might- vanish-under-trump/

Depository Library Council. (1993). Alternatives for Restructuring the Depository Library Program: A Report to the Superintendent of Documents and the Public Printer from the Depository Library Council. *Administrative Notes*, Vol. 14, No. 13 (p18-54). https://

www.govinfo.gov/app/details/GOVPUB-
GP3-6d5f969d656100f7baf75767b99ec09c

Depository Library Council. (2004). Incentives Document: Progress
Report: How's the Carrot Crop Doing? *Administrative Notes, 25*(11)
14-22. https://www.govinfo.gov/app/details/GOVPUB-GP3-
beba6f0831ee27bc68bb775d3fd9b1f8

Derrida, J. (1996). *Archive fever : a Freudian impression* (E. Prenowitz,
Trans.). University of Chicago Press Chicago.

Digital Preservation Coalition. (2023). *Bit List 2023: The Global List of
Endangered Digital Species* (4th ed.). https://doi.org/10.7207/
dpcbitlist-23

Digital.gov. (2019). An introduction to accessibility. https://digital.gov/
resources/introduction-accessibility/

digital.gov. (2023). U.S. Web Design System (USWDS). https://
designsystem.digital.gov/

Dobreski, B. (2021). Descriptive Cataloging: The History and Practice of
Describing Library Resources. *Cataloging & Classification Quarterly,
59*(2-3), 225–241. https://doi.org/10.1080/01639374.2020.1864693

DOI Foundation. (2023). Multiple DOI Resolution. In *DOI Handbook.*
https://www.doi.org/doi-handbook/HTML/multiple-doi-
resolution.html

Dooley, C., & Thomas, G. (2019). The Library of Congress Web Archives:
Dipping a Toe in a Lake of Data. *The Signal, Library of Congress blog.*
blogs.loc.gov/thesignal/2019/01/the-library-of-congress-web-archives-
dipping-a-toe-in-a-lake-of-data

Douneva, M., & Jaron, R. (2016). Effects of Different Website Designs on
First Impressions, Aesthetic Judgements and Memory Performance
after Short Presentation. *Interacting with Computers.* https://doi.org/
10.1093/iwc/iwv033

Doyle, C. D. (2003). Federal Electronic Information. In *Encyclopedia of
Library and Information Science* (pp. 1068-1078).

Draper, D., & Lederer, N. (2013). Analysis of Readex's Serial Set MARC
records: Improving the data for the library catalog. *Government*

Information Quarterly, *30*(1), 87–98. https://doi.org/10.1016/
j.giq.2012.06.010

Dunsire, G., Hillmann, D., & Phipps, J. (2017). Reconsidering universal
bibliographic control in light of the semantic web Functional Future
for Bibliographic Control (pp. 112-124). Routledge. https://
ecommons.cornell.edu/server/api/core/bitstreams/09e4f331-
aaf2-4791-a80f-1833e4ceaf08/content

Elings, M. W., & Waibel, G. (2007). Metadata for all: Descriptive
standards and metadata sharing across libraries, archives and
museums. *First Monday*, *12*(3). https://doi.org/10.5210/fm.v12i3.1628

Eliot, T. S. (1945). The Man of Letters and the Future of Europe. *The
Sewanee Review*, *53*(3), 333–342. http://www.jstor.org/stable/
27537594

End of Term Web Archive. (2004). Project Background, End of Term Web
Archive. https://web.archive.org/web/20240309223757/http://
eotarchive.cdlib.org/background.html

Environmental Data and Governance Initiative. (2017). *EDGI*. https://
envirodatagov.org/

FDLP. (2011). *CD-ROM Equivalence Project*. https://web.archive.org/
web/20120612222257/fdlp.gov/home/about/917-cdrom-equivalent-
project

FDLP. (2016). Library Services and Content Management Performance
Metrics. https://web.archive.org/web/20210705023239/https://
www.fdlp.gov/file-repository/about-the-fdlp/lscm-performance-
metrics/2479-cumulative-metrics-fy2005-to-present/file

FDLP. (2021a). *Web Archiving*. https://web.archive.org/web/
20210415053340/https://www.fdlp.gov/377-projects-active/2020-
web-archiving#what_federal_web

FDLP. (2022a). All or Mostly Online Federal Depository Libraries. https://
www.fdlp.gov/guidance/all-or-mostly-online-federal-depository-
libraries

FDLP. (2022b). *Agency Subscription Databases.* https://www.fdlp.gov/requirements-and-guidance/collections-and-databases/article-agency-subscription

FDLP. (2023a). *Item Number System.* https://www.fdlp.gov/guidance/item-number-system

FDLP. (2023b). List of Current Partnerships (2023). https://web.archive.org/web/20230831044904/https://www.fdlp.gov/sites/default/files/file_repo/Partnerships-2023MAY26.xlsx

FDLP. (2023c). *Web Archiving.* FDLP Projects. https://www.fdlp.gov/project-list/web-archiving

FDLP. (2025a). *GPO Partnerships.* https://fdlp.gov/collaborations-with-gpo/partnerships

FDLP. (2025b). *Welcome to the FDLP Resource Guides.* https://libguides.fdlp.gov/?b=g&d=a

FDLP. (2021b). *The National Collection of U.S. Government Public Information.* https://www.fdlp.gov/about-the-fdlp/the-national-collection

Fearon, D., Gunia, B., Pralle, B., Lake, S., & Sallans, A. (2013). ARL SPEC Kit 334: Research data management services. https://libraopen.lib.virginia.edu/downloads/gf06g2710

Federal Agencies Digital Guidelines Initiative. (2023). FADGI Program: Impacts and Benefits. https://www.digitizationguidelines.gov/about/FADGI-impacts_20230329.pdf

Federal Depository Library Council. (2007). Depository Library Conference [transcript] April 15, 2007. *Christopher Boone, Digital Reporter.* https://web.archive.org/web/20150907193126/https://www.fdlp.gov/file-repository/outreach/events/depository-library-council-dlc-meetings/meeting-minutes-and-transcripts/690-transcripts-from-the-spring-2007-dlc-meeting/file

Federal Documents Task Force of the ALA Government Documents Round Table. (1973). Suggestions to GPO. A Letter to the Superintendent of Documents. *Documents to the People, 1*(3), 21–

28. https://babel.hathitrust.org/cgi/pt?
id=inu.30000003627985&seq=7

Federal Judicial Center, Federal Judicial History Office. (2010). Guide to Research in Federal Judicial History. https://purl.fdlp.gov/GPO/gpo18787

Fenton, E. G. (2011). Protecting Future Access Now: Developing a Prototype Preservation Model for Digital Books. https://works.hcommons.org/records/pqgb1-ptc66

Ferguson, A. W., & Kehoe, K. (1993). Access vs. ownership: What is most cost effective in the sciences. *Journal of Library Administration, 19*(2), 89. https://search.ebscohost.com/login.aspx?direct=true&db=lxh&AN=9410111025&site=ehost-live

Ferriero, D. S. (2017). NARA's Role under the Presidential Records Act and the Federal Records Act. *Prologue, 49*(2). https://www.archives.gov/publications/prologue/2017/summer/archivist-pra-fra

Free Government Information. (2025). FGI Library. https://freegovinfo.info/librarypage

Fretwell, L. (2024). *Metadata, Open Graph and government websites.* https://govfresh.com/thoughts/metadata-open-graph-government-websites

Furner, J. (2015). Information science is neither. *Library trends, 63*(3), 362–377. https://doi.org/10.1353/lib.2015.0009

Gade, E. K., & Wilkerson, J. (2017). The .GOV internet archive: a big data resource for political science. *The Political Methodologist, 16.* https://faculty.washington.edu/jwilker/559/GovPaper.pdf

Gaffney, M., & Massie, D. (2022). Putting a Price on Sharing: Fresh Cost Data as an Essential Ingredient of Evaluating Interlibrary Loan Services. *Library Trends, 70*(3), 409–429. https://doi.org/10.1353/lib.2022.0002

Gano, G., & Linden, J. (2007). Government Information in Legacy Formats. *D-Lib Magazine, 13*(7/8). https://doi.org/10.1045/july2007-linden

Gapen, D. K. (1989). Statement On Behalf Of The Association Of Research Libraries. In *Federal information dissemination policies and practices : hearings* (pp. 338-366). https://catalog.hathitrust.org/Record/007610170

GCN Staff. (2001). NARA's Web archive plan irks agencies. *GCN.* https://web.archive.org/web/20220123125910/https://gcn.com/2001/02/naras-web-archive-plan-irks-agencies/276414/

Gehrke, G., Beck, M., Paz, A., Wilson, A., Nost, E., & Poudrier, G. (2021). Access Denied: Federal Web Governance Under the Trump Administration. https://envirodatagov.org/publication/access-denied-federal-web-governance-under-the-trump-administration/

Gerken, J. L. (2004). A Librarian's Guide to Unpublished Judicial Opinions. *Law Library Journal, 96*(3), 475–501. https://web.archive.org/web/20101119122335/http://aallnet.org/products/pub_llj_v96n03/2004-28.pdf

Ghasemisharif, M., Snyder, P., Aucinas, A., & Livshits, B. (2019). Speedreader: Reader mode made fast and private. In *The World Wide Web Conference* (pp. 526-537). https://arxiv.org/pdf/1811.03661

Giesecke, J. (2011). Finding the Right Metaphor: Restructuring, Realigning, and Repackaging Today's Research Libraries. *Journal of Library Administration, 51*(1), 54 – 65. https://search.ebscohost.com/login.aspx?direct=true&db=lxh&AN=57225161&site=ehost-live

Gilliland, A. (2016). Setting the Stage. In M. Baca (Ed.), *Introduction to Metadata: Pathways to Digital Information*. Getty Information Institute. https://www.getty.edu/publications/intrometadata/setting-the-stage/

Ginsberg, W. (2015). Common Questions About Federal Records and Related Agency Requirements. *Congressional Research Service, R43072.* https://crsreports.congress.gov/product/pdf/download/R/R43072/R43072.pdf/

Gleim, L. C., & Decker, S. (2020). *Timestamped URLs as persistent identifiers*. In (pp. 11-16). https://mepdaw-ws.github.io/2020/accepted-papers/MEPDaW_2020_paper_8.pdf

GODORT. (1991). GODORT's Principles on Government Information. *Documents to the People, 19*(1), 12–14. https://stacks.stanford.edu/file/xj550cr3472/xj550cr3472.pdf

GODORT Ad Hoc Committee On The Internet. (1996). Government Information In The Electronic Environment (A GODORT Whitepaper) January 1996. *Documents to the People, 24*(1), 21–39. https://babel.hathitrust.org/cgi/pt?id=mdp.39015082961536&seq=9

GODORT Committee on Legislation. (1990). GODORT Information Principles. *Documents to the people, 18*(1), 36–37. https://stacks.stanford.edu/file/tf887vb1386/tf887vb1386.pdf

GODORT Committee on Legislation. (1996). Resolution On The Principles For Federal Government Information. *Documents to the People, 24*(3), 164. https://stacks.stanford.edu/file/fk624sx0092/fk624sx0092.pdf

GODORT Committee on Legislation. (2004). *Key Principles On Government Information.* https://www.ala.org/rt/godort/2004-resolutions#5

GODORT Federal Documents Task Force. (1992). Priorities for Disseminating Electronic Products and Service from the U.S. Government Printing Office. *Documents to the People, 20*(2), 78–79. https://babel.hathitrust.org/cgi/pt?id=mdp.39015064547337&seq=5

govdoc-l. (2000). *[Census discussions].* https://lists.psu.edu/cgi-bin/wa?A1=200009-200009&L=GOVDOC-L&s=1000&O=&D=&TOC=&S=

GPO. (1994). *GPO ACCESS Status Report [1994].* https://web.archive.org/web/20090510081147/http://www.gpoaccess.gov//biennial/1994/report.pdf

GPO. (1995a). *Catalog of U.S. Government Publications (CGP).* https://catalog.gpo.gov/

GPO. (1995b). The Electronic Federal Depository Library Program: Transition Plan, FY 1996 – FY 1998. *Administrative Notes, 16*(18). https://purl.fdlp.gov/GPO/gpo141434

GPO. (1996a). Federal Depository Library Program: Information Dissemination And Access Strategic Plan, FY 1996 - FY 2001. In

GPO's Study to Identify Measures Necessary for a Successful Transition to a More Electronic Federal Depository Library Program. https://purl.fdlp.gov/GPO/LPS4220

GPO. (1996b). *Study to identify measures necessary for a successful transition to a more electronic Federal Depository Library Program: as required by Legislative Branch Appropriations Act, 1996, Public Law 104-53.* U.S. Government Printing Office. https://purl.fdlp.gov/GPO/LPS4220

GPO. (2004a). Collection of Last Resort. https://web.archive.org/web/20041022165420if_/http://www.gpoaccess.gov/about/reports/clr0604draft.pdf

GPO. (2004b). Concept of Operations (ConOps V1.0) for the Future Digital System Final. https://web.archive.org/web/20041101064957/www.gpo.gov/news/2004/ConOps_1004.pdf

GPO. (2004c). A Strategic Vision for the 21st Century. https://web.archive.org/web/20151121034646/http://www.fdlp.gov/file-repository/about-the-fdlp/strategic-plan-for-the-fdlp/38-a-strategic-vision-for-the-21st-century/file

GPO. (2004d). Digital Archiving through Bibliographic Control Begins in November. *Administrative Notes, 25*(12), 6. https://web.archive.org/web/20170102164733/https://www.fdlp.gov/file-repository/historical-publications/administrative-notes/2004-adnotes/894-administrative-notes-vol-25-no-12-13/file

GPO. (2005a). Web Harvesting. *Administrative Notes, 26*(1), 8–9. https://www.govinfo.gov/content/pkg/GOVPUB-GP3-44720622961f099e8c0fb4b6d5bc6171/pdf/GOVPUB-GP3-44720622961f099e8c0fb4b6d5bc6171.pdf

GPO. (2005b). Harvesting Federal Digital Publications for GPO's Information Dissemination (ID) Programs. *Information Dissemination Policy Statement ID 73, Administrative Notes v.26 no.07(7/15/05).* https://www.fdlp.gov/file-repository/historical-publications/administrative-notes/2005-adnotes/880-administrative-notes-vol-26-no-07/file

GPO. (2006a). Web Discovery And Harvesting. *Update for ALA January 2006.* https://web.archive.org/web/20160917013956/https://www.fdlp.gov/file-repository/outreach/events/gpo-attended-events/2006-gpo-attended-events/523-american-library-association-ala-update-midwinter-2006/file

GPO. (2006b). GPO's Future Digital System: System Releases And Capabilities. *Version 4.0.* https://www.govinfo.gov/media/FDsys_RD_v4.0.pdf

GPO. (2007a). *Government Information Locator Service (GILS): About.* https://web.archive.org/web/20070714013038/http://www.gpoaccess.gov/gils/about.html

GPO. (2007b). Managing Web Harvested Content. *Depository Library Council Meeting Proceedings, 2007 Spring, Denver Colorado, Council Session - Web Harvesting.* https://web.archive.org/web/20160917024343/https://www.fdlp.gov/file-repository/outreach/events/depository-library-council-dlc-meetings/2007-meeting-proceedings/spring-dlc-meeting-denver-colorado/311-council-session-web-harvesting/file

GPO. (2009). Federal Depository Library Program Strategic Plan, 2009 - 2014: Creating an informed citizenry and improving the quality of life: Draft Discussion Document. https://web.archive.org/web/20090528141715/http://www.fdlp.gov/home/repository/doc_download/37-fdlp-strategic-plan-2009-2014-draft-4172009

GPO. (2011a). GPO's Strategic Plan FY 2011-2015. https://www.gpo.gov/docs/default-source/mission-vision-and-goals-pdfs/gpo-strategic-plan-fy-2011-2015.pdf

GPO. (2011b). Oyez! Oyez! Federal Court Opinions in FDsys. https://web.archive.org/web/20160917014921/http://www.fdlp.gov/all-newsletters/featured-articles/1144-oyez-oyez-federal-court-opinions-in-fdsys

GPO. (2015a). 2014 Annual Report. https://www.govinfo.gov/app/details/GOVPUB-GP-92b29e12d4da029bf0c0a3cc5cf710c8

GPO. (2015b). *Publishing Guidelines from GPO.* https://www.gpo.gov/
docs/default-source/forms-standards-pdf-files/
gpo_publishing_guidelines.pdf?sfvrsn=2

GPO. (2015c). Electronic Code of Federal Regulations (e-CFR) XML User
Guide. https://www.govinfo.gov/bulkdata/ECFR/resources/ECFR-XML-
User-Guide.pdf

GPO. (2016a). *GPO Style Manual.* https://www.govinfo.gov/content/pkg/
GPO-STYLEMANUAL-2016/pdf/GPO-STYLEMANUAL-2016.pdf

GPO. (2016b). Keeping America Informed: The U.S. Government
Publishing Office, A Legacy of Service to the Nation, 1861-2016.
Revised Edition. https://www.govinfo.gov/content/pkg/GPO-
KEEPINGAMERICAINFORMED-2016/pdf/GPO-
KEEPINGAMERICAINFORMED-2016.pdf

GPO. (2016c). National Plan for Access to U.S. Government
Information: A Framework for a User-centric Service Approach To
Permanent Public Access. https://web.archive.org/web/
20160505183934/https://www.fdlp.gov/file-repository/about-the-fdlp/
gpo-projects/national-plan-for-access-to-u-s-government-information/
2700-national-plan-for-access-to-u-s-government-information-a-
framework-for-a-user-centric-service-approach-to-permanent-public-
access/file

GPO. (2016d). Scope of Government Information Products Included in
the Cataloging and Indexing Program and Disseminated Through the
Federal Depository Library Program. *Effective: 02/05/2008,
Superintendent Of Documents Public Policy Statement 2016-1.*
https://web.archive.org/web/20161226014949/https://www.fdlp.gov/
file-repository/about-the-fdlp/policies/superintendent-of-documents-
public-policies/2739-scope-of-government-information-products-
included-in-the-cataloging-and-indexing-program-and-disseminated-
through-the-federal-depository-library-program/file

GPO. (2016e). GPO's System of Online Access: Collection Development
Plan. https://web.archive.org/web/20161216234604/https://
www.fdlp.gov/file-repository/about-the-fdlp/gpo-projects/trustworthy-

digital-reports/2812-gpo-s-system-of-online-access-collection-development-plan/file

GPO. (2016f). Harvesting Digital Federal Government Information Dissemination Products for GPO's Superintendent of Documents Programs. *Public Policy Statement, 2016-5*. https://www.fdlp.gov/file-download/download/public/2687

GPO. (2017). Responses to Questions for the Record Submitted by the Committee on House Administration. https://www.gpo.gov/docs/default-source/congressional-relations-pdf-files/testimonies/qfr-cha-september-2017.pdf

GPO. (2018a). *Legal Requirements & Program Regulations of the Federal Depository Library Program*. https://www.fdlp.gov/file-download/download/public/17694

GPO. (2018b). *Comments on Draft Legislation to Amend Title 44, USC* https://www.gpo.gov/docs/default-source/congressional-relations-pdf-files/testimonies/title-44---gpo-comments-on-dec-version-of-draft-title-44-bill---jan-2018---rev.pdf

GPO. (2018c). *New govinfo API*. https://www.govinfo.gov/features/api

GPO. (2019). Access to U.S. Government Information within Scope of the Public Information Programs of the Superintendent of Documents. *Public Policy Statement, 2019-3*. https://www.fdlp.gov/file-download/download/public/16459

GPO. (2020a). GPO Director Advocates for Modernizing Congressional Document Formats. *Press Release, 20-09*. https://www.gpo.gov/who-we-are/news-media/news-and-press-releases/gpo-director-advocates-for-modernizing-congressional-document-formats

GPO. (2020b). Government Publishing Office Legislative Proposals. https://web.archive.org/web/20210301170859/https://www.fdlp.gov/file-repository/about-the-fdlp/title-44-revision/4696-government-publishing-office-legislative-proposals-july-15-2020/file

GPO. (2020c). Disseminating Information Products to the Public through GPO's Federal Depository Library Program. *Circular Letter, No. 1056*. https://www.gpo.gov/how-to-work-with-us/agency/circular-letters/

disseminating-information-products-to-the-public-through-gpo-s-federal-depository-library-program-aug-2020

GPO. (2021a). *Federal Depository Library Program Web Archive*. https://archive-it.org/home/FDLPwebarchive

GPO. (2021b). GPO Annual Report 2020. https://www.gpo.gov/docs/default-source/news-content-pdf-files/2020_annualreport.pdf

GPO. (2021c). *What's Available [GOVINFO]*. https://www.govinfo.gov/help/whats-available

GPO. (2021d). GPO And Libraries Set Goal To Make Every U.S. Government Document Accessible. *News Release, No. 21-07*. https://www.gpo.gov/docs/default-source/news-content-pdf-files/2021/gpo-and-libraries-set-goal-to-make-every-u-s-government-document-accessible.pdf

GPO. (2021e). *Release Notes [govinfo]*. https://www.govinfo.gov/features/september-2021-release-notes

GPO. (2021f). GPO Digitizes List Of Publications The Federal Government Has Produced Since The 1800S. *Press Release, No. 21-16*. https://www.gpo.gov/docs/default-source/news-content-pdf-files/2021/gpo-digitizes-list-of-publications-the-federal-government-has-produced-since-the-1800s.pdf

GPO. (2022a). *What's Included in Search Results*. https://www.govinfo.gov/help/search-results

GPO. (2022b). Updates On USLM Projects and XPub. https://usgpo.github.io/innovation/resources/CDTF20220621/CDTF-GPO-USLM-XPub-20220621.pdf

GPO. (2023a). CGP on GitHub. https://github.com/usgpo/cataloging-records

GPO. (2023b). Federal Publishing Council. https://www.gpo.gov/how-to-work-with-us/agency/federal-publishing-council

GPO. (2023c). *Preservation Agreement with NARA*. https://www.govinfo.gov/about/policies#nara-preservation

GPO. (2023d). Services for Agencies. https://www.gpo.gov/how-to-work-with-us/agency/services-for-agencies

GPO. (2023e). *Government Publishing Office Legislative Proposals for Title 44, Chapters 17 and 19 (March 2023) [Discussion Draft].* https://www.fdlp.gov/sites/default/files/file_repo/2023-Leg-Proposals-FDLP-Sales.pdf

GPO. (2025). *GAO Reports and Comptroller General Decisions.* https://www.govinfo.gov/app/collection/gaoreports

GPO. (2025b). xpub. https://github.com/usgpo/xpub

GPO. (2025c). Federal Depository Library Directory, Library of Congress. https://ask.gpo.gov/s/account/001t000000T9HJMAA3/serial-and-government-publications-division

GPO. (2010). *Government Information Locator Service (GILS): Main/Search Page.* https://web.archive.org/web/20100505081658/http://www.gpoaccess.gov/gils/index.html

GPO Library Programs Service. (1995). Recommended Minimum Technical Guidelines for Federal Depository Libraries. *Administrative Notes, 16*(2), 5–7. https://www.govinfo.gov/content/pkg/GOVPUB-GP3-b02fa99174bbd220d5b2e4e2cdef6d7d/pdf/GOVPUB-GP3-b02fa99174bbd220d5b2e4e2cdef6d7d.pdf

GPO Library Programs Service. (1998). *Managing the FDLP Electronic Collection: A Policy and Planning Document.* U.S. Government Printing Office. https://www.govinfo.gov/content/pkg/GOVPUB-GP3-0e48a21dc042b081e4681f699262e525/pdf/GOVPUB-GP3-0e48a21dc042b081e4681f699262e525.pdf

GPO Superintendent of Documents. (1995). The Electronic Federal Depository Library Program: Transition Plan, FY 1996 – FY 1998 [superseded]. https://web.archive.org/web/20090326004054/http://www.gpo.gov/su_docs/fdlp/retired/transit.html

GPO Superintendent Of Documents. (2020). Government Publications Authorized for Discard by Regional Depository Libraries. *Effective: 10/01/2020, Superintendent Of Documents Public Policy Statement 2020-1.* https://web.archive.org/web/20201202150836/https://

www.fdlp.gov/file-repository/about-the-fdlp/policies/superintendent-of-documents-public-policies/4400-government-publications-authorized-for-discard-by-regional-depository-libraries-2/file

GPO Superintendent Of Documents. (1983). Format of Publications Distributed to Depository Libraries, Superintendent Of Documents SOD 13. *Administrative Notes, 8*(21). https://www.govinfo.gov/content/pkg/GOVPUB-GP3-098d39274632b02f9ad70bc7ce3f54b8/pdf/GOVPUB-GP3-098d39274632b02f9ad70bc7ce3f54b8.pdf

GPO Superintendent Of Documents. (2001). *Dissemination/Distribution Policy for the Federal Depository Library Program.* Superintendent Of Documents Policy Statement, SOD 71. https://web.archive.org/web/20010410062230/http://www.access.gpo.gov/su_docs/fdlp/pubs/sod71.html

GPO Superintendent Of Documents. (2005). *Dissemination/Distribution Policy for the Federal Depository Library Program.* Superintendent Of Documents Policy Statement ID 71 EFFECTIVE DATE: June 21, 2005. https://web.archive.org/web/20060210093750/http://www.access.gpo.gov/su_docs/fdlp/pubs/policies/id71_06-21-05.pdf

GPO Superintendent Of Documents. (2006). *Dissemination/Distribution Policy for the Federal Depository Library Program.* Superintendent Of Documents Policy Statement SOD 301 EFFECTIVE DATE: June 1, 2006. https://web.archive.org/web/20060709174245/http://www.access.gpo.gov/su_docs/fdlp/coll-dev/sod301.pdf

GPO Superintendent Of Documents. (2008). *Superintendent Of Documents Public Policy Statement 2016-1, Scope Of Government Information Products Included In The Cataloging And Indexing Program And Disseminated Through The Federal Depository Library Program, Effective: 02/05/2008.* https://web.archive.org/web/20161219120147/https://www.fdlp.gov/file-repository/about-the-fdlp/policies/superintendent-of-documents-public-policies/2739-scope-of-government-information-products-included-in-the-cataloging-and-indexing-program-and-disseminated-through-the-federal-depository-library-program/file

GPO Superintendent Of Documents. (2018). *Guidelines for Establishing Shared Regional Depository Libraries*. https://www.fdlp.gov/guidelines-for-establishing-shared-regional-depository-libraries-sod-guidance-document-2018-1-1

GPO Superintendent Of Documents. (2019a). Scope of Government Information Products Included in the Cataloging and Indexing Program and Disseminated Through the Federal Depository Library Program. *Superintendent Of Documents Public Policy Statement 2019-1*. https://web.archive.org/web/20190413052948/https://www.fdlp.gov/file-repository/about-the-fdlp/policies/superintendent-of-documents-public-policies/3875-scope-of-government-information-products-included-in-the-cataloging-and-indexing/file

GPO Superintendent Of Documents. (2019b). *Superintendent Of Documents Public Policy Statement 2019-1 Scope of Government Information Products Included in the Cataloging and Indexing Program and Disseminated Through the Federal Depository Library Program, Effective: 04/10/2019, Supersedes SOD-PPS-2016-1.* https://www.fdlp.gov/file-download/download/public/15592

GPO, Library Services & Content Management. (2015). LSCM year in review. https://www.govinfo.gov/content/pkg/GOVPUB-GP3-43cf83018708cd97a8145f815825a0d4/pdf/GOVPUB-GP3-43cf83018708cd97a8145f815825a0d4.pdf

GPO, Library Services & Content Management. (2024). FY 2023 LSCM Year in Review. https://www.govinfo.gov/app/details/GOVPUB-GP3-a50d912cea2e959b64dc8c122f812c41/summary

GPO, Library Services and Content Management. (2011). FY2011 Year In Review. https://web.archive.org/web/20151025040731/https://www.fdlp.gov/file-repository/about-the-fdlp/lscm-year-in-review/2116-lscm-fy2011-year-in-review/file

GPO. Office of Inspector General. (2017). Additional Information Needed for Ensuring Availability of Government Information Through the Federal Depository Library Program. *Audit Report, 18-01.* https://www.gpo.gov/docs/default-source/inspector-general/audits/2018/18-01.pdf

Granger, S. (2002). Digital Preservation and Deep Infrastructure. *D-Lib Magazine, 8*(2). http://www.dlib.org/dlib/february02/granger/02granger.html

Greenberg, J. (2018). Metadata and Digital Information [ELIS Classic]. In J. D. McDonald & M. Levine-Clark (Eds.), *Encyclopedia of Library and Information Science* (4th ed.). CRC Press. https://doi.org/10.1081/E-ELIS4

Greenstein, D. I. (2004). *Access in the Future Tense.* Council on Library and Information Resources. https://www.clir.org/wp-content/uploads/sites/6/pub126.pdf

Greenwood, J. M., & Bockweg, G. (2012). Insights to building a successful e-filing case management service: US Federal Court experience. *International Journal for Court Administration, 4.* https://iacajournal.org/articles/74/files/submission/proof/74-1-164-1-10-20131023.pdf

Griffith, C. (2015). Why Do We Still Have Unpublished Opinions? *Forbes.* https://www.forbes.com/sites/taxanalysts/2015/11/05/why-do-we-still-have-unpublished-opinions/

Griffith, J. B., & Relyea, H. (1996). Compilation of Statutes Authorizing Dissemination of Government Information to the Public. *CRS-1996-OSS-0001.*

Grotke, A. (2011). Web Archiving at the Library of Congress. *Computers in Libraries, 31*(10). https://www.infotoday.com/cilmag/dec11/Grotke.shtml

GSA. (2021). Site scanning documentation, terms. https://github.com/GSA/site-scanning-documentation/blob/main/pages/terms.md

GSA. (2022). *govt-urls, 2_govt_urls_federal_only.csv.* https://github.com/GSA/govt-urls/blob/master/2_govt_urls_federal_only.csv

GSA. (2022b). Government Domains Outside .Gov and .Mil. https://search.gov/about/policy/govt-urls.html

GSA. (2023). Past scanning efforts. https://github.com/GSA/site-scanning-documentation/blob/main/about/project-management/project-history.md

GSA, Digital.gov. (2020). *Guide to the Site Scanning program.* https://digital.gov/guides/site-scanning/

GSA, Digital.gov. (2023). Guide to the Site Scanning program, Technical details. https://digital.gov/guides/site-scanning/technical-details/

Guerrini, M. (2023). *From Cataloguing to Metadata Creation : A Cultural and Methodological Introduction.* Facet Publishing. http://ebookcentral.proquest.com/lib/ucsd/detail.action?docID=7109478

Halpern, H. N. (2022). Prepared Statement before the Subcommittee on Legislative Branch Committee on Appropriations U.S. House of Representatives United States Government Publishing Office FY 2023 Budget Hearing. https://www.gpo.gov/docs/default-source/congressional-relations-pdf-files/testimonies/gpo-director-halpern-hearing-subcommittee--legislative-branch-appropriations-4-27-2022.pdf

Harari, Y. N. (2024). *Nexus: a brief history of information networks from the Stone Age to AI.* Random House. https://search.worldcat.org/en/title/1439595634

Harper, R., Thereska, E., Lindley, S., Banks, R., Gosset, P., Odom, W., Smyth, G., & Whitworth, E. (2011). What is a File? *Microsoft Research Ltd. Technical Report, MSR-TR-2011-109.* https://www.microsoft.com/en-us/research/wp-content/uploads/2011/10/MSR-TR-2011-109.pdf

Harrington, W. G. (1984). A Brief History of Computer-Assisted Legal Research. *Law Library Journal, 77*(3), 543–556. https://heinonline.org/HOL/LandingPage?handle=hein.journals/llj77&div=45&id=&page=

Harris, S. (2001). Agencies frustrated by order to save Web pages. *GovExec.* https://www.govexec.com/technology/2001/02/agencies-frustrated-by-order-to-save-web-pages/8483/

Harvard Library Innovation Lab Team. (2025). Announcing the Data.gov Archive. https://lil.law.harvard.edu/blog/2025/02/06/announcing-data-gov-archive/

Harvard Library. (2023). *Locations & Hours.* https://library.harvard.edu/libraries

Harvey, R., & Weatherburn, J. (2018). *Preserving Digital Materials* (3rd ed.). Rowman and Littlefield. https://search.worldcat.org/en/title/8163834253

Heanue, A. (1981). *Less access to less information by and about the U.S. government.* https://freegovinfo.info/less_access

Henry, L. (1998). Schellenberg in cyberspace. *The American Archivist, 61*(2), 309–327. https://doi.org/10.17723/aarc.61.2.f493110467x38701

Hernandez, J., & Byrnes, T. (2004). CD-ROM Analysis Project. *(presentation at Spring 2004 Depository Library Council meeting, St. Louis, MO, April 21, 2004); unpublished PowerPoint.*

Hernon, P., & Relyea, H. C. (1995). Government Publishing: Past to Present. *Government Information Quarterly, 12*(3), 309–330. https://doi.org/10.1016/0740-624X(95)90021-7

Hernon, P. (1986). The management of United States government information resources: An assessment of OMB Circular A-130. *Government Information Quarterly, 3*(3), 279–290. https://www.sciencedirect.com/science/article/abs/pii/0740624X86900808

Hernon, P., & Saunders, L. (2009). The Federal Depository Library Program in 2023: One Perspective on the Transition to the Future. *College & Research Libraries, 70*(4), 351–370. http://crl.acrl.org/index.php/crl/article/download/16021/17467

Hershowitz, A. (2023). Legislative Modernization: Where we are today. *LinkedLegislation Blog.* https://blog.linkedlegislation.com/2023-03-09-legislative-modernization-today/

Heslop, H., Davis, S., & Wilson, A. (2002). *An approach to the preservation of digital records.* National Archives of Australia. https://web.archive.org/web/20210119060805/https://www.ltu.se/cms_fs/1.83844!/file/An_approach_Preservation_dig_records.pdf

Hirtle, P. B. (2001). Government Records in a Digital World [1998]. In D. B. Marcum (Ed.), *Development of Digital Libraries : An American Perspective.* Greenwood Publishing Group, Incorporated. http://ebookcentral.proquest.com/lib/ucsd/detail.action?docID=3000707

Hirtle, P. B. (2008). The history and current state of digital preservation in the United States. In E. L. Westbrooks & K. Jenkins (Eds.), *Metadata and Digital Collections: A Festschrift in Honor of Tom Turner*. Cornell University Library. https://hdl.handle.net/1813/45862

Holdsworth, D. (1996). Medium is NOT the Message OR Indefinitely Long-Term File Storage at Leeds University. *NASA publication 3340, Proceedings of the 5th NASA Goddard Conference on Mass Storage Systems and Technologies*. https://web.archive.org/web/19970725191056/http://esdis.gsfc.nasa.gov/msst/conf1996/A6_07Holdsworth.html

Holterhoff, S. (1990a). Report on GPO Documents Collection and Legislative Archives. *Administrative Notes, 11*(3). https://www.govinfo.gov/app/details/GOVPUB-GP3-5555e63eb2b69eacff86df6bad64f878

Holterhoff, S. (1990b). Summary, Fall Meeting Depository Library Council, Arlington, Virginia October 18-20, 1989. *Administrative Notes, 11*(3), 15. https://www.govinfo.gov/app/details/GOVPUB-GP3-5555e63eb2b69eacff86df6bad64f878

House Committee on Appropriations. (2011). Legislative Branch Appropriations Bill, 2012. *House Report 112–148*. https://www.govinfo.gov/content/pkg/CRPT-112hrpt148/pdf/CRPT-112hrpt148.pdf

House Office of the Legislative Counsel [U.S.]. (2023). *Quick Guide to Legislative Drafting*. https://legcounsel.house.gov/sites/evo-subsites/legcounsel.house.gov/files/documents/quick_guide.pdf

Housewright, R., & Schonfeld, R. C. (2011). Modeling a Sustainable Future for the United States Federal Depository Library Program's Network of Libraries in the 21st Century: Final Report of Ithaka S+R to the Government Printing Office. https://web.archive.org/web/20160916221757/https://www.fdlp.gov/file-repository/about-the-fdlp/gpo-projects/future-direction-of-the-fdlp-modeling-initiative/2004-final-report-with-recommendations-and-accompanying-gpo-statement/file

Hughes, S. (2019). The Federal Courts Are Running An Online Scam. *Politico*. https://www.politico.com/magazine/story/2019/03/20/pacer-court-records-225821/

Hull, T. J. (1998). Business statistics and the national economy: Electronic records of the Department of Commerce in the National Archives and Records Administration. *Business and Economic History*, 257–263. https://www.jstor.org/stable/23703088

Hyland, B., & Wood, D. (2011). The Joy of Data: A Cookbook for Publishing Linked Government Data on the Web. In D. Wood (Ed.), *Linking Government Data* (pp. 3-26). Springer. https://link.springer.com/chapter/10.1007/978-1-4614-1767-5_1

ICPSR. (2025). *DataLumos*. https://archive.icpsr.umich.edu/datalumos/home

ICPSR. (2025b). Inter-university Consortium for Political and Social Research. https://www.icpsr.umich.edu/

IFLA Study Group on the Functional Requirements for Bibliographic Records. (2009). Functional Requirements For Bibliographic Records Final Report. https://repository.ifla.org/handle/123456789/811

Interagency Committee on Government Information Electronic Records Policy Working Group. (2004). *Recommendations For The Effective Management Of Government Information On The Internet And Other Electronic Records* (Vol. Submitted to the Director of the Office of Management and Budget and the Archivist of the United States). https://web.archive.org/web/20050210142215/http://www.cio.gov/documents/ICGI/ICGI-207e-report.pdf

Interagency Committee on Government Information Web Content Management Workgroup. (2004). *Recommended Policies and Guidelines for Federal Public Websites, Final Report* (Vol. Submitted to The Office of Management and Budget). https://web.archive.org/web/20041222162900if_/http://www.cio.gov:80/documents/ICGI/ICGI-June9report.pdf

International Telecommunication Union. (1992). *Information Technology – Digital Compression And Coding Of Continuous-Tone Still Images –*

Requirements And Guidelines. Recommendation. https://www.w3.org/Graphics/JPEG/itu-t81.pdf

Internet Archive. (2022). *CDX File Format.* http://www.archive.org/web/researcher/cdx_file_format.php

Internet Archive. ([n.d.]). *Access Services.* https://webservices.archive.org/pages/access

Internet Assigned Numbers Authority. (2024). Media Types. https://www.iana.org/assignments/media-types/media-types.xhtml

Invest in Open Infrastructure. (2022). *Catalog of Open Infrastructure Services (COIs)*. https://investinopen.org/catalog/

ISO. (2003). *ISO 14721:2003 Open archival information system - Reference model [Withdrawn]* (1st ed.). https://www.iso.org/standard/24683.html

ISO. (2012a). *ISO 16363:2012 Audit and certification of trustworthy digital repositories* (1st ed.). https://www.iso.org/standard/56510.html

ISO. (2012b). *ISO 14721:2012 Open archival information system (OAIS) - Reference model* (2nd ed.). https://www.iso.org/standard/57284.html

ISO. (2023). *Information technology : Metadata registries (MDR) : Part 1 : Framework* (4 ed.). ISO/IEC. https://www.iso.org/standard/78914.html

Jacobs, J. A. (1990). U.S. Government Computer Bulletin Boards: A Modest Proposal For Reform. *Government Publications Review, 17*(5), 393–396. https://doi.org/10.1016/0277-9390(90)90048-I

Jacobs, J. A. (2000). Re: "Letter to Depository Library Directors from Fran Buckley". *govdoc-l.* https://lists.psu.edu/cgi-bin/wa?A2=GOVDOC-L;4cef32e.00&S=a

Jacobs, J. A. (2002). *Access to Government Information, pre and post 9/11.* In *Intellectual freedom in libraries post 9/11.* https://freegovinfo.info/node/7877/

Jacobs, J. A. (2009a). Federal Depository Library Program: Services and Collections. *Against the Grain, 21*(2). https://doi.org/10.7771/2380-176X.2554

Jacobs, J. A. (2011). Privatization of GPO, Defunding of FDsys, and the Future of the FDLP. *Free Government Information (FGI)*. http://freegovinfo.info/node/3416

Jacobs, J. A. (2013). When we depend on pointing instead of collecting. *Free Government Information (FGI)*. http://freegovinfo.info/node/3900

Jacobs, J. A. (2014). Born-digital us federal government information: Preservation and access. *Prepared For The Center For Research Libraries Global Resources Collections Forum, Leviathan: Libraries and Government Information in the Age of Big Data*. https://live-crl-www.pantheonsite.io/sites/default/files/d6/attachments/pages/Leviathan%20Jacobs%20Report%20CRL%20%C6%92%20%283%29.pdf

Jacobs, J. A. (2015). The FDLP Historical Collections. *Free Government Information (FGI)*. http://freegovinfo.info/node/10142

Jacobs, J. A., & Jacobs, J. R. (2017). Digital Deposit And The Biennial Survey: context and actions. *Free Government Information*. https://freegovinfo.info/node/12457/

Jacobs, J. A., & Jacobs, J. R. (2019). Digital Deposit, Good for All: Vision, Myths, Reality. *Free Government Information*. https://freegovinfo.info/node/13361/

Jacobs, J. A., & Jacobs, J. R. (2021). Government recommendations to preserve government information not preserved by government. *Free Government Information*. https://freegovinfo.info/node/14266/

Jacobs, J. A., & Jacobs, J. R. (2025). The government information crisis is bigger than you think it is. *Free Government Information*. https://freegovinfo.info/node/14747/

Jacobs, J. A., & Jacobs, J. R. (2016). Looking Backward, Looking Forward. *Free Government Information*. http://freegovinfo.info/node/10752

Jacobs, J. A., & Jacobs, J. R. (2013). The Digital-Surrogate Seal of Approval: a Consumer-oriented Standard. *D-Lib Magazine, 19*(3/4). https://doi.org/10.1045/march2013-jacobs

Jacobs, J. R. (2009b). Critical GPO systems and the FDLP cloud. https://freegovinfo.info/node/2704/

Jacobs, J. R. (2016). Less Access to Less Information By and About the U.S. Government. *Free Government Information.* https://freegovinfo.info/less_access2

Jacobs, J. R. (2017). "Issued for Gratuitous Distribution": The History of Fugitive Documents and the FDLP. *Against the Grain, 29*(6). https://purl.stanford.edu/yc376vd9668

Jacobs, J. R. (2021). Great news! GPO announces "fugitive documents" will now be called "unreported publications". *Free Government Information.* https://freegovinfo.info/node/14173/

James, B. (2003). Depository Library Council Minutes, Information Exchange, Oct. 21, 2003. https://web.archive.org/web/20150907193254if_/http://www.fdlp.gov/file-repository/outreach/events/depository-library-council-dlc-meetings/meeting-minutes-and-transcripts/610-dlc-meeting-minutes-2000-2003/file

Johnson, E., & Kubas, A. (2018). Spotlight on Digital Government Information Preservation: Examining the Context, Outcomes, Limitations, and Successes of the DataRefuge Movement. *In the Library with the Lead Pipe.* https://www.inthelibrarywiththeleadpipe.org/2018/information-preservation/

Jones, M. (2002). The Cedars Project. *Library and Information Research News, 26*(84), 11–16. http://eprints.rclis.org/6045/1/article84a.pdf

Joulin, A., Grave, E., Bojanowski, P., & Mikolov, T. (2016). Bag of Tricks for Efficient Text Classification. *arXiv:1607.01759 [cs].* http://arxiv.org/abs/1607.01759

Judicial Conference of the United States. (1963). Reports of the proceedings of the Judicial Conference of the United States. https://catalog.gpo.gov/F/?func=direct&doc_number=000326760&local_base=GPO01PUB

Kahle, B. (2022). What is the Democracy's Library? *Internet Archive Blog.* https://blog.archive.org/2022/11/30/what-is-the-democracys-library/

Keele, D. (2012). District Court Data Sources: Implications and Opportunities. *Law & Courts Newsletter Of The Law & Courts Section Of The American Political Science Association, 22*(2), 15–20. https://citeseerx.ist.psu.edu/document?repid=rep1&type=pdf&doi=1dad002d4b3acbe05c2e3c799b26a7a11d26793b#page=15

Keenan, T. M., Burroughs, J. M., & Ebanues, S. (2013). Partners in Collaborative Cataloging: The U.S. Government Printing Office and the University of Montana. *University of Montana, Cataloging & Classification Quarterly, 51*(1-3), 118–128. https://scholarworks.umt.edu/cgi/viewcontent.cgi?&article=1006&context=ml_pubs

Kessler, R. R. (1996). A brief history of the Federal Depository Library Program: A personal perspective. *Journal of Government Information, 23*(4), 369–380. https://doi.org/10.1016/1352-0237(96)00017-2

Khan, H., Caruso, B., Corson-Rikert, J., Dietrich, D., Lowe, B., & Steinhart, G. (2011). DataStaR: Using the Semantic Web approach for Data Curation. *International Journal of Digital Curation, 62*(2). https://doi.org/10.2218/ijdc.v6i2.197

Kraemer, D., & Dahlen, A. (2018). Intricacies of Digital Content and PURLs. *Federal Depository Library Conference.* https://web.archive.org/web/20181014091457if_/https://www.fdlp.gov/file-repository/outreach/events/depository-library-council-dlc-meetings/2018-meeting-proceedings-1/2018-depository-library-conference-1/3649-intricacies-of-purls-and-linking-to-digital-content-slides/file

Kram, L. (1998). Why continue to be a depository library if it is all on the internet anyway. *Government Information Quarterly, 15*(1), 57–71. https://doi.org/10.1016/S0740-624X(98)90015-6

Kurtz, M. J. (2005). *Challenges and Changes: Electronic Records Management.* In *RACO 2005.* https://www.archives.gov/files/records-mgmt/presentations/kurtz-raco2005.ppt

Lamdan, S. (2017). Lessons from DataRescue: The Limitations of Grassroots Climate Change Data Preservation and the Need for Federal Records Law Reform. *U. Pa. L. Rev. Online, 166*, 231. https://scholarship.law.upenn.edu/cgi/viewcontent.cgi?article=1214&context=penn_law_review_online

Landgraf, M. R., Zwaard, K., Haun-Mohamed, R., & Mauldin, J. (2010). Digital Preservation Framework, Goals, and Challenges at the U. S. Government Printing Office. In *Proceedings of the 2010 Roadmap for Digital Preservation Interoperability Framework Workshop.* New York, NY, USA Gaithersburg, Maryland, USA: Association for Computing Machinery. https://dl.acm.org/doi/abs/10.1145/2039274.2039280

Lavoie, B. (2014). *The Open Archival Information0 System (OAIS) Reference Model: Introductory Guide* (2nd ed.). https://www.dpconline.org/docs/technology-watch-reports/1359-dpctw14-02/file

Lavoie, B., & Dempsey, L. (2004). Thirteen Ways of Looking at Digital Preservation. *D-Lib Magazine, 10*(7/8). https://doi.org/10.1045/july2004-lavoie

Lavoie, B. F. (2003). *The Incentives to Preserve Digital Materials: Roles, Scenarios, and Economic Decision-Making.* White paper. https://www.oclc.org/content/dam/research/activities/digipres/incentives-dp.pdf

Lavoie, B. F., Malpas, C., & Shipengrover, J. D. (2012). Print Management at "Mega-scale": a Regional Perspective on Print Book Collections in North America. http://www.oclc.org/research/publications/library/2012/2012-05.pdf

Lazorchak, B. (2012). All In! Embedded Files in PDF/A. *The Signal.* https://blogs.loc.gov/thesignal/2012/11/all-in-embedded-files-in-pdfa/

Legal Information Institute. (2025). *Our Work: Collections.* https://about.law.cornell.edu/our-work/

Legislative Branch Innovation Hub. (2019). Data Standardization. https://usgpo.github.io/innovation/data_standardization/

Levy, D. M. (1994). *Fixed or fluid? Document stability and new media* (pp. 24-31). https://dl.acm.org/doi/pdf/10.1145/192757.192760

Library of Congress Working Group on the Future of Bibliographic Control. (2008). On the Record: Report of The Library of Congress Working Group on the Future of Bibliographic Control. http://www.loc.gov/bibliographic-future/news/lcwg-ontherecord-jan08-final.pdf

Lifton, R. J. (1961). *Thought reform and the psychology of totalism : a study of "brainwashing" in China.* W.W. Norton.

Lindley, S. E., Smyth, G., Corish, R., Loukianov, A., Golembewski, M., Luger, E. A., & Sellen, A. (2018). Exploring new metaphors for a networked world through the file biography. In *Proceedings of the 2018 CHI Conference* (pp. 1-12). https://researchonline.rca.ac.uk/3055/1/paper118.pdf

Link, F. E., Tosaka, Y., & Weng, C. (2015). Mining and Analyzing Circulation and ILL Data for Informed Collection Development. *College & Research Libraries, 76*(6), 740–755. https://doi.org/10.5860/crl.76.6.740

Lippincott, S. K. (2018). Environmental Scan of Government Information and Data Preservation Efforts and Challenges. https://perma.cc/X5XG-ZFFD

LOCKSS Program. (2022). Digital Federal Depository Library Program. https://www.lockss.org/join-lockss/networks/digital-federal-depository-library-program

Lofgren, Z. (2021). *Social Media Review: Members Of The U.S. House Of Representatives Who Voted To Overturn The 2020 Presidential Election.* https://lofgren.house.gov/socialreview

Love, J. (1993). SEC'S EDGAR On Net, What Happened And Why. *TAP-INFO Internet Distribution List.* https://web.archive.org/web/20060719053309/http://parallel.park.org/Cdrom/TheNot/Mail/EdgarUp/msg00088.html

Lucier, R. (1996). The University as Library. *Follett Lecture Series, University of Leeds*. https://www.ukoln.ac.uk/services/papers/follett/lucier/paper.html

Lynch, C. (1999). Canonicalization: A Fundamental Tool to Facilitate Preservation and Management of Digital Information. *D-Lib Magazine*, 5(9). https://doi.org/10.1045/september99-lynch

Lynch, C. (2000). From Automation to Transformation: Forty Years of Libraries and Information. *Educause Review*, 60–68. https://www.cni.org/wp-content/uploads/2000/02/auto-2-transform.pdf

Lynch, C. A. (2003). Institutional Repositories: Essential Infrastructure For Scholarship In The Digital Age. *portal: Libraries and the Academy*, 3(2), 327–336. http://muse.jhu.edu/journals/portal_libraries_and_the_academy/v003/3.2lynch.html

MacGilvray, D. R. (2006). A Short History of GPO. *GPO*. https://www.govinfo.gov/content/pkg/GPO-AShortHistoryofGPO/text-submitted/GPO-AShortHistoryofGPO.txt

MacGilvray, M. W., & Walters, J. M. (1995). Electronic Capabilities of Federal Depository Libraries, Summer 1994. https://purl.fdlp.gov/GPO/gpo68277

Madison, J. (1910). Letter to W.T. Barry, August 4, 1822. In G. Hunt (Ed.), *The Writings of James Madison Vol. IX 1819-1836* (p. 103). G.P. Putnum's Sons.

Malone, C. (2016). How Trump's White House Could Mess With Government Data. *FiveThirtyEight*. https://fivethirtyeight.com/features/how-trumps-white-house-could-mess-with-government-data/

Malpas, C. (2011). Cloud-sourcing Research Collections: Managing Print in the Mass-digitized Library Environment. *OCLC Research*. https://www.oclc.org/content/dam/research/publications/library/2011/2011-01.pdf

Maniatis, P., Roussopoulos, M., Giuli, T. J., Rosenthal, D. S. H., & Baker, M. (2005). The LOCKSS peer-to-peer digital preservation system. *ACM Trans. Comput. Syst.*, 23(1), 2–50. https://doi.org/10.1145/1047915.1047917

Mann Library. (2025). *USDA Economics, Statistics and Market Information System.* https://usda.library.cornell.edu/about?locale=en

Maret, S. (2016). *On Their Own Terms: A Lexicon with an Emphasis on Information-Related Terms Produced by the U.S. Federal Government* (6th ed.). https://sgp.fas.org/library/maret.pdf

Marks, J. (2011). Kundra names task force to consolidate federal websites. *Nextgov.* https://www.nextgov.com/technology-news/2011/07/kundra-names-task-force-to-consolidate-federal-websites/49388/

Marks, J. (2013). Dot-govs Dip Below 1,000. https://www.nextgov.com/emerging-tech/2013/05/there-are-now-fewer-1000-government-web-domains/63158/

Martin, P. W. (2008). Online Access to Court Records: From Documents to Data, Particulars to Patterns. *Villanova Law Review, 53*(5), 855–888. https://digitalcommons.law.villanova.edu/cgi/viewcontent.cgi?article=1125&context=vlr

Martin, P. W. (2018). District Court Opinions That Remain Hidden Despite a Long-Standing Congressional Mandate of Transparency - the Result of Judicial Autonomy and Systemic Indifference. *Law Library Journal, 110*(3), 305–332. https://www.aallnet.org/wp-content/uploads/2018/10/LLJ_110n3_01_martin.pdf

Martin, R. S. (2003). Cooperation and Change: Archives, libraries and museums in the United States. In *69th IFLA General Conference and Council.* https://web.archive.org/web/20180528201528/http://archive.ifla.org/IV/ifla69/papers/066e-Martin.pdf/

Mayernik, M. (2011). The distributions of MARC fields in bibliographic records. *Library Resources \& Technical Services, 54*(1), 40–54. https://journals.ala.org/lrts/article/download/5051/6116

McCreadie, M., & Rice, R. E. (1999). Trends in analyzing access to information. Part I: cross-disciplinary conceptualizations of access. *Information Processing & Management, 35*(1), 45–76. https://doi.org/10.1016/S0306-4573(98)00037-5

McLaughlin, P. A., & Stover, W. (2021). Drafting X2RL: A Semantic Regulatory Machine-Readable Format. *MIT Computational Law Report*. https://law.mit.edu/pub/draftingx2rl/release/2

Minick, C. (2011). On PACER and FDsys. *Justia Law Blog*. https://lawblog.justia.com/2011/05/13/on-pacer-and-fdsys/

Moreland, J. L. (2021). Is Open Access Equal Access? *Documents to the People*, *49*(3/4), 42–48. https://journals.ala.org/index.php/dttp/issue/viewIssue/817/586

Morrissey, S., Meyer, J., Bhattarai, S., Kurdikar, S., Ling, J., Stoeffler, M., & Thanneeru, U. (2010). Portico: A case study in the use of xml for the long-term preservation of digital artifacts. In *Proceedings of the International Symposium on XML for the Long Haul: Issues in the Long-term Preservation of XML*. Montreal, CA. https://www.balisage.net/Proceedings/vol6/print/Morrissey01/BalisageVol6-Morrissey01.html

Mossoff, A. (2020). Radical OSTP Proposal Would Undermine American Research and Sacrifice American Intellectual Property. *The Heritage Foundation, Legal Memorandum No. 263*. https://www.heritage.org/sites/default/files/2020-05/LM264.pdf

Mulligan, S. J. (2025). Inside the race to archive the US government's websites. *MIT Technology Review*. https://www.technologyreview.com/2025/02/07/1111328/inside-the-race-to-archive-the-us-governments-websites/

Mumma, C. C. (2023). *The Future of Digital Preservation – BPE 2023 Plenary Talk*. https://www.tdl.org/2023/06/the-future-of-digital-preservation-bpe-2023-plenary-talk/

Musurlian, P. (2024). NARA's looming digitization deadline for agencies means the end of paper. *Federal News Network*. https://federalnewsnetwork.com/federal-newscast/2024/05/naras-looming-digitization-deadline-for-agencies-means-the-end-of-paper/

Nabe, J., & Fowler, D. C. (2015). Leaving the "Big Deal" … Five Years Later. *The Serials Librarian*, *69*(1), 20–28. https://doi.org/10.1080/0361526x.2015.1048037

Naisbitt, J. (1982). *Megatrends : ten new directions transforming our lives*. Warner Books New York.

NARA. (2017). Guidance on Presidential Records from the National Archives and Records Administration. https://purl.fdlp.gov/GPO/gpo127666

NARA. ([n.d.]). *Record Group Clusters and Locations, Judicial*. https://www.archives.gov/research/alic/tools/record-group-clusters.html#jud

NARA. (2005a). *Bulletin 2006-02 NARA Guidance for Implementing Section 207(e) of the E-Government Act of 2002 [superseded]*. https://www.archives.gov/records-mgmt/bulletins/2006/2006-02.html

NARA. (1990). Electronic Records Management: Final Rule. *Federal Register, 55*(89), 19216–19221. https://www.govinfo.gov/content/pkg/FR-1990-05-08/pdf/FR-1990-05-08.pdf

NARA. (1995). Electronic Mail Systems, Final Rule. *Federal Register, 60*(166), 44633–44642. https://www.govinfo.gov/app/details/FR-1995-08-28

NARA. (1998). *Press Release, National Archives and Records Administration Holds Final Meeting of Electronic Records Work Group*. Press Release. https://www.archives.gov/press/press-releases/1998/nr98-137.html

NARA. (2001). *NWM 05.2001- Snapshot of Agency Public Web Sites, Memorandum to Agency Records Officers and Information Resource Managers*. https://www.archives.gov/records-mgmt/faqs/agencies.html

NARA. (2004a). *Initiatives: Web Content Guidance [superseded]*. https://www.archives.gov/records-mgmt/initiatives/web-content-records.html

NARA. (2004b). *Bulletin 2005-02 Harvest of Agency Public Web Sites [superseded]*. https://www.archives.gov/records-mgmt/bulletins/2005/2005-02.html

NARA. (2005b). *2004 Presidential Term Web Harvest*. https://web.archive.org/web/20050519020336/http://www.webharvest.gov/collections/peth04/

NARA. (2005c). *Electronic Records Archive, About ERA*. https://web.archive.org/web/20051023125147/http://www.archives.gov/era/about/index.html#background

NARA. (2005d). NARA Guidance on Managing Web Records. https://www.archives.gov/files/records-mgmt/pdf/managing-web-records-index.pdf

NARA. (2006). Implications of Recent Web Technologies for NARA Web Guidance [no longer current]. https://www.archives.gov/records-mgmt/initiatives/web-tech.html

NARA. (2008a). NWM 13.2008 Memorandum To Federal Agency Contacts: End-of-Administration web snapshot [withdrawn]. https://web.archive.org/web/20080917133157/http://archives.gov/records-mgmt/memos/nwm13-2008.html

NARA. (2008b). *Web Harvest Background Information (NWM 13.2008 web snapshot reply) [superseded]*. https://web.archive.org/web/20080916030003/https://www.archives.gov/records-mgmt/memos/nwm13-2008-brief.html

NARA. (2008c). *Web Harvest Background Information [PDF version of "NWM 13.2008 web snapshot reply"] [superseded]*. https://www.archives.gov/files/records-mgmt/pdf/nwm13-2008-brief.pdf

NARA. (2010). A Report on Federal Web 2.0 Use and Record Value. https://www.archives.gov/files/records-mgmt/resources/web2.0-use.pdf

NARA. (2013). Request for Records Disposition Authority, 2001 Web Snapshot Initiative, DAA-0064-2013-0001. https://www.archives.gov/files/records-mgmt/rcs/schedules/independent-agencies/rg-0064/daa-0064-2013-0001_sf115.pdf

NARA. (2016). *Publications of the U.S. Government*. https://www.archives.gov/research/guide-fed-records/groups/287.html

NARA. (2018). Federal Agency Records Management 2017 Annual Report. https://www.archives.gov/files/records-mgmt/resources/2017-farm-annual-report.pdf

NARA. (2019a). *Government Publications and Library Materials*. https://www.archives.gov/research/start/government-pubs.html

NARA. (2019b). *FAQs for GRS 6.2, Federal Advisory Committee Records*. https://www.archives.gov/records-mgmt/grs/faqs-for-grs-6-2

NARA. (2020a). *Publications of the U.S. Government (Record Group 287)*. https://www.archives.gov/legislative/research/gov-pubs.html

NARA. (2020b). *National Archives Frequently Asked Questions*. https://www.archives.gov/faqs

NARA. (2021). Government Publications, 1861-1992. https://catalog.archives.gov/id/306711

NARA. (2021a). *Archived Presidential White House Websites*. https://www.archives.gov/presidential-libraries/archived-websites

NARA. (2021b). *Frequently Asked Questions of Individual GRS Schedules*. https://www.archives.gov/files/records-mgmt/grs/faqs-individual-grs.pdf

NARA. (2021c). *Congressional Web Harvest*. https://www.archives.gov/legislative/research/web-harvest.html

NARA. (2021d). [Fiscal Year 2022 Budget Request] FY 2022 Congressional Justification. https://www.archives.gov/files/about/plans-reports/performance-budget/2022performance-budget.pdf

NARA. (2022a). 2022–2026 Strategic Plan. https://www.archives.gov/files/about/plans-reports/strategic-plan/nara-2022-2026-strategic-plan-march-2022.pdf

NARA. (2022b). *Advisory Committee On The Records Of Congress*. https://www.archives.gov/legislative/cla/advisory-committee

NARA. (2022c). *The Center for Legislative Archives*. https://www.archives.gov/legislative

NARA. (2024). Metadata Requirements for Permanent Electronic Records. https://www.archives.gov/records-mgmt/policy/metadata-compiled

NARA. (2025). Presidential Term 2004 [web harvest]. https://www.webharvest.gov/collections/peth04/

NARA. (2008d). *NARA Bulletin 2008-03 Scheduling existing electronic records (NARA Bulletin 2006-02) [superseded]*. https://www.archives.gov/records-mgmt/bulletins/2008/2008-03.html

NARA Center for Legislative Archives. (2017). Rules of Access. https://www.archives.gov/legislative/research/rules-of-access.html

NARA Office of General Counsel. (2016). *Basic Laws and Authorities of the National Archives and Records Administration*. https://www.archives.gov/files/about/laws/basic-laws-book-2016.pdf?_ga=2.172885478.432275793.1624560716-87961582.1624560716

NARA. Electronic Records Work Group. (1998). Report to the Archivist of the United States. https://web.archive.org/web/19981202013640/http://www.nara.gov/records/grs20/reprt914.html

NASA. (2023). NRRS 1441.1 NASA Records Retention Schedules. https://nodis3.gsfc.nasa.gov/NPR_attachments/NRRS_1441.1.pdf

National Academy Of Public Administration. (2013). Rebooting The Government Printing Office: Keeping America Informed in the Digital Age. *Academy Project Number: 2170*. https://napawash.org/academy-studies/rebooting-the-government-printing-office-keeping-america-informed-in-the-di

National Digital Information Infrastructure and Preservation Program (U.S.). (2002). *Preserving our digital heritage : plan for the National Digital Information Infrastructure and Preservation Program : a collaborative initiative of the Library of Congress*. https://purl.fdlp.gov/GPO/LPS27275

National Digital Information Infrastructure and Preservation Program (U.S.). (2018). *What is Digital Preservation?* https://web.archive.org/web/20180204154645/http://www.digitalpreservation.gov/about/

Nelson, T. H. (1974). *Computer Lib / Dream Machines*. [self publishished]. https://search.worldcat.org/en/title/217227165

Neto, A. J. R., Borges, M. M., & Roque, L. (2017). *Preliminary Study About the Applicability of a Service-Oriented Architecture in the OAIS Model Implementation*. New York, NY: Association for Computing Machinery. https://dx.doi.org/10.1145/3144826.3145381

Neubert, M. (2015). Introducing the Federal Web Archiving Working Group. *The Signal*. https://blogs.loc.gov/thesignal/2015/02/introducing-the-federal-web-archiving-working-group/

Nost, E., Gehrke, G., Poudrier, G., Lemelin, A., Beck, M., Wylie, S., & et al. (2021). Visualizing changes to US federal environmental agency websites, 2016–2020. *PLoS ONE, 16*(2). https://doi.org/10.1371/journal.pone.0246450

NTIS. (2022). Limited Access Death Master File. https://www.ntis.gov/ladmf/ladmf.xhtml

Nuclear Waste Technical Review Board. (2014). Request for Records Disposition Authority [Approved]. *DAA-0220-2014-0013*. https://www.archives.gov/files/records-mgmt/rcs/schedules/independent-agencies/rg-0220/daa-0220-2014-0013_sf115.pdf

OCLC. (2003). *Digital Archive*. https://web.archive.org/web/20031009115438/http://www.oclc.org/digitalarchive/

OCLC Online Computer Library Center Incorporated. (1999). *Welcome to OCLC*. https://web.archive.org/web/19990129003941/http://www.oclc.org/oclc/menu/t-home1.htm

Office of Technology Assessment. (1988). *Informing the nation : federal information dissemination in an electronic age.* U.S. Congress, Office of Technology Assessment : For sale by the Supt. of Docs., U.S. G.P.O. https://purl.fdlp.gov/GPO/LPS27552

OMB. (1993). Federal Information Resources Management (Circular A-130), Revision; Notice. *Federal Register, 58*(126). https://www.govinfo.gov/content/pkg/FR-1993-07-02/pdf/FR-1993-07-02.pdf

OMB. (2004). Policies for Federal Agency Public Websites. *Memorandum, M-05-04*. https://georgewbush-whitehouse.archives.gov/omb/memoranda/fy2005/m05-04.pdf

OMB. (2011). Implementing Executive Order 13571 on Streamlining Service Delivery and Improving Customer Service. MEMORANDUM, M 11-24. https://obamawhitehouse.archives.gov/sites/default/files/omb/memoranda/2011/m11-24.pdf

OMB. (2016). *Circular No. A-130: Managing Information as a Strategic Resource.* https://bidenwhitehouse.archives.gov/wp-content/uploads/legacy_drupal_files/omb/circulars/A130/a130revised.pdf

OMB. (1985). Circular No. A-130: Management of Federal Information Resources; Final Publication. *Federal Register, 50*(247), 52730–52751. https://www.govinfo.gov/app/details/FR-1985-12-24

Otlet, P. (1990). The Science Of Bibliography And Documentation [1903]. In W. B. Rayward (Ed.), *International organisation and dissemination of knowledge : selected essays of Paul Otlet* (pp. 71-86). Elsevier. https://search.worldcat.org/en/title/21875536

Ozbekhan, H. (1968). The Triumph of Technology: 'Can' Implies 'Ought'. https://doi.org/https://doi.org/10.4324/9781351106450

Pasek, J. E. (2017). Historical Development and Key Issues of Data Management Plan Requirements for National Science Foundation Grants: A Review. *Issues in Science and Technology Librarianship, 87*(Summer). https://doi.org/10.29173/istl1709

Payne, L. (2007). Library storage facilities and the future of print collections in North America. https://www.oclc.org/content/dam/research/publications/library/2007/2007-01.pdf

PEGI Project. (2024). *Charting a FAIR Direction for the US Government Information Ecosystem.* https://www.pegiproject.org/blog/fairgovinfo

Pennock, M. (2013). Web-Archiving. *DPC Technology Watch Report 13-01.* https://doi.org/10.7207/twr13-01

Petersen, R. E. (2017). Statement. In *Hearing on "Transforming GPO for the 21st Century and Beyond: Part 4".* Congressional Research Service. http://docs.house.gov/meetings/HA/HA00/20171011/106494/HHRG-115-HA00-Wstate-PetersenE-20171011.pdf

Peterson, K., & Jacobs, J. A. (2005). *Government Information in the Digital Era: Free Culture or Controlled Substance?* In *Symposium on Free Culture and the Digital Library.* Atlanta, GA. https://digital.library.unt.edu/ark:/67531/metadc97947/m2/1/high_res_d/FCDL-Proceedings-FINAL_0.pdf

Petrich, M.-E., & Becker, S. (2024). Happy 16th Birthday to the USDocs PLN! *LOCKSS*. https://www.lockss.org/news/happy-16th-birthday-usdocs-pln

Philipson, J. (2019). Identifying PIDs playing FAIR. *Data Science, 2*(1-2), 229–244. https://content.iospress.com/articles/data-science/ds190024

Phillips, M. E. (2016). How many of the EOT2008 PDF files were harvested in EOT2012. *mark e. phillips journal*. https://vphill.com/journal/post/5872/

Planets Preservation and Long-term Access through NETworked Services. (2007). About Planets. https://planets-project.eu/about/

Pomerantz, J. (2015). *Metadata*. MIT Press. https://doi.org/10.7551/mitpress/10237.001.0001

Powell, V. (2001). NARA's new view on Web snapshots. *FCW*. https://fcw.com/workforce/2001/04/naras-new-view-on-web-snapshots/217547/

Project On Government Oversight. (2008). *POGO and other groups urge National Archives to preserve website records of the federal government*. https://web.archive.org/web/20181017020955/https://www.pogo.org/letter/2008/04/pogo-and-other-groups-urge-national-archives-to-preserve-website-records-of-federal-government/

PTAB Primary Trustworthy Digital Repository Authorisation Body Ltd. (2018). *United States Government Publishing Office*. PTAB-TDRMS 0002. http://www.iso16363.org/iso-certification/certified-clients/united-states-government-publishing-office/

Public Environmental Data Partners. (2025). Data + Screening Tools. https://screening-tools.com/about

Qasim, U., Davis, C., Garnett, A., Marks, S., & Moosberger, M. (2018). Research Data Preservation in Canada : A White Paper. https://open.library.ubc.ca/collections/52387/items/1.0371946

Raymond, N. (2021). Free PACER? *Reuters*. https://www.reuters.com/legal/legalindustry/free-pacer-bill-end-fees-online-court-records-advances-senate-2021-12-09/

Reilly, B. F. (2008). Center for Research Libraries' Auditing and Certification of Digital Archives. *The Charleston Advisor, 9*(3), 59–60. https://annurev.publisher.ingentaconnect.com/content/annurev/tca/2008/00000009/00000003/art00019

Relyea, H. C. (2004). *Public Printing Reform: Issues and Actions* (Vol. CRS report 98-687 GOV). Congressional Research Service. https://digital.library.unt.edu/ark:/67531/metadc819141/m2/1/high_res_d/98-687_2004May27.pdf

Research Data Alliance. (2024). *Metadata Standards Catalog Index of subjects*. https://rdamsc.bath.ac.uk/subject-index

Research Libraries Group. (1997). *The Research Libraries Group, Inc.* https://web.archive.org/web/19970503091101/http://www.rlg.org/

Research Libraries Group. (2001). Attributes of a Trusted Digital Repository: Meeting the Needs of Research Resources An RLG-OCLC Report DRAFT FOR PUBLIC COMMENT. https://citeseerx.ist.psu.edu/document?repid=rep1&type=pdf&doi=738ed14c54a6b05523a9539cebfd6076919b685a

Research Libraries Group. (2002). Trusted Digital Repositories: Attributes and Responsibilities: An RLG-OCLC Report. https://web.archive.org/web/20090916084206/http://www.oclc.org/programs/ourwork/past/trustedrep/repositories.pdf

Richardson Jr, J. V., Frisch, D. C. W., & Hall, C. M. (1980). Bibliographic Organization Of U.S. Federal Depository Collections. *Government Publications Review, 7*(6), 463–480. https://doi.org/10.1016/0196-335X(80)90026-6

Rieger, O. Y. (2018). The State of Digital Preservation in 2018, A Snapshot of Challenges and Gaps. *Ithaka S+R Issue Brief.* https://doi.org/10.18665/sr.310626

Rieger, O. Y., Schonfeld, R. C., & Sweeney, L. (2022). The Effectiveness and Durability of Digital Preservation and Curation Systems. https://doi.org/10.18665/sr.316990

Riva, P., Le Bœuf, P., Žumer, M., & IFLA FRBR Review Group
   Consolidation Editorial Group. (2017). IFLA Library Reference
   Model : a conceptual model for bibliographic information. https://
   www.ifla.org/files/assets/cataloguing/frbr-lrm/ifla-lrm-august-2017.pdf

RLG-NARA Task Force on Digital Repository Certification. (2007).
   *Trustworthy Repositories Audit & Certification (TRAC): Criteria and
   Checklist Version 1.0*. Center for Research Libraries. https://live-crl-
   www.pantheonsite.io/sites/default/files/d6/attachments/pages/
   trac_0.pdf

Rowberry, S. (2023). Historiographies of Hypertext. In *Proceedings of the
   34th ACM Conference on Hypertext and Social Media* (pp. 1-10).
   https://dl.acm.org/doi/pdf/10.1145/3603163.3609038

Rowe, J. S. (1992). Review: Management, preservation and access for
   electronic records with enduring value: Response to
   recommendations: National Archives and Records Administration
   [NARA's typescript response to above report]. *Government
   Publications Review, 19*(3), 304–306. https://doi.org/
   10.1016/0277-9390(92)90080-U

Rubin, R. E., & Rubin, R. G. (2020). *Foundations of Library and
   Information Science* (5th ed.). Facet Publishing. https://
   search.worldcat.org/en/title/1138996906

Rumsey, A. S. (2011). *But Storage is Cheap: Digital Preservation in the
   Age of Abundance*. Preservation Lecture Series. https://
   concerningelectronicrecordmanagement.blogspot.com/2011/07/but-
   storage-is-cheap-digital.html

Russell, J. C. (2010). Challenges and Opportunities for Federal
   Depository Libraries in the Digital Age. *Against the Grain, 22*(5).
   https://doi.org/10.7771/2380-176X.5644

Safdar, M., Ashiq, M., & Rehman, S. U. (2022). Research data services in
   libraries: a systematic literature review. *Information Discovery and
   Delivery, 51*(2), 151–165. https://doi.org/10.1108/IDD-04-2021-0044

Sansobrino, J. C., Wood, K. A., & Kramer, M. E. (1993). A Guide To
   Accessing Collections Of Major U.S. Government Publications In The

Library Of Congress. https://web.archive.org/web/20010417041148/
https://www.loc.gov/rr/news/us.govt.pubs.html

Schellenberg, T. R. (1956). Informational Values. In *The Appraisal of Modern Records*. https://www.archives.gov/research/alic/reference/
archives-resources/appraisal-informational-values.html

Schiller, H. I. (1991). Public information goes corporate. *Library Journal*,
116(16), 42–45. https://eric.ed.gov/?id=EJ434751

Schonfeld, R. C., & Housewright, R. (2009a). Documents for a Digital
Democracy: A Model for the Federal Depository Library Program in
the 21st Century. http://www.ithaka.org/ithaka-s-r/research/
documents-for-a-digital-democracy/
Documents%20for%20a%20Digital%20Democracy.pdf

Schonfeld, R. C., & Housewright, R. (2009b). What to Withdraw: Print
Collections Management in the Wake of Digitization. 28. http://
www.sr.ithaka.org/research-publications/what-withdraw-print-
collections-management-wake-digitization

Schultz, M. (2010). *MetaArchive Cooperative TRAC Audit Checklist*.
https://web.archive.org/web/20220120055531/https://
metaarchive.org/wp-content/uploads/2017/03/ma_trac_audit.pdf

Schultze, S. J. (2018). The price of ignorance: The constitutional cost of
fees for access to electronic public court records. *Georgetown Law
Journal*, *106*(4), 1197–1227. https://www.law.georgetown.edu/
georgetown-law-journal/wp-content/uploads/sites/26/2018/07/The-
Price-of-Ignorance.pdf

Schwartz, J. (2009). An Effort to Upgrade a Court Archive System to Free
and Easy. *New York Times*. https://www.nytimes.com/2009/02/13/us/
13records.html

Schwarzkopf, L. C. (1978). The GPO Microform Program: Its History And
Status. *Documents to the People*, 6(4). https://purl.stanford.edu/
vh394rv6107

Search.gov. (2020). *Metadata and tags you should include in your
website*. https://search.gov/indexing/metadata.html

Securities and Exchange Commission. (2019). Request for Records Disposition Authority [Modified Approved Version]. *DAA-0266-2013-0002*. https://www.archives.gov/files/records-mgmt/rcs/schedules/independent-agencies/rg-0266/daa-0266-2013-0002_sf115.pdf

Senate Committee on Governmental Affairs. (2002). *E-government Act of 2001: report of the Committee on Governmental Affairs United States Senate to accompany S. 803* (Vol. Senate Report 107-174). U.S. Government Printing Office. http://hdl.handle.net/2027/pur1.32754073720678

Shah, U. U., & Gul, S. (2019). LOCKSS, CLOCKSS & PORTICO: A look into digital preservation policies. *Library philosophy and practice, 2841*. https://digitalcommons.unl.edu/libphilprac/2481/

Shaw, T. S. (1966). Library Associations and Public Documents. *Library Trends*, 167–177. https://www.ideals.illinois.edu/items/6233

Shendruk, A., & Rampell, C. (2025). How Trump is reshaping reality by hiding data [Opinion]. *Washington Post*. https://www.washingtonpost.com/opinions/interactive/2025/government-data-trump-deletion/

Shepherd, E. (2018). Archival Science. In J. D. McDonald & M. Levine-Clark (Eds.), *Encyclopedia of Library and Information Sciences* (4th ed.). CRC Press. https://doi.org/10.1081/E-ELIS4

Shuler, J. (2008). 60 Days to Government Information Liberation. *Free Government Information (FGI)*. https://freegovinfo.info/node/2163/

Skinner, K., & Halbert, M. (2009). The MetaArchive Cooperative: A Collaborative Approach to Distributed Digital Preservation. *Library Trends, 57*(3), 371–392. https://doi.org/10.1353/lib.0.0042

Sleeman, B. (2009). This Page Intentionally Blank: Writing the Next Chapter in the Future of the Federal Depository Library Program. *Occasional Paper Series*. https://digitalcommons.law.umaryland.edu/cgi/viewcontent.cgi?article=1523&context=fac_pubs

Software Preservation Network. (2024). *Emulation-as-a-Service Infrastructure*. https://www.softwarepreservationnetwork.org/emulation-as-a-service-infrastructure/

Soni, J., & Goodman, R. (2017). A Man in a Hurry: Claude Shannon's New York Years. *IEEE Spectrum: Technology, Engineering, and Science News*. https://spectrum.ieee.org/geek-life/history/a-man-in-a-hurry-claude-shannons-new-york-years

Spence, J. (2006). Preserving the cultural heritage. *Aslib Proceedings, 58*(6), 513–524. https://doi.org/10.1108/00012530610713597

Springshare. (2023). LibGuides & LibGuides CMS. https://springshare.com/libguides/

Star, S. L., & Ruhleder, K. (1996). Steps toward an ecology of infrastructure: design and access for large information spaces. *Information Systems Research, 7*(1), 111–134. https://doi.org/10.1287/isre.7.1.111

Steinhart, G., Dietrich, D., & Green, A. (2009). Establishing Trust in a Chain of Preservation: The TRAC Checklist Applied to a Data Staging Repository (DataStaR). *D-Lib, 15*(9/10). http://www.dlib.org/dlib/september09/steinhart/09steinhart.html

Sternstein, A. (2005). GPO awaits Web-harvesting technology. *Federal Computer Week*. https://web.archive.org/web/20210506075331/https://fcw.com/articles/2005/11/07/gpo-awaits-webharvesting-technology.aspx

Stiglitz, Joseph E., Orszag, Peter R. Orszag, Jonathan M., Computer & Communications Industry Association. (2000). *The Role of Government in a Digital Age*. https://www.ccianet.org/wp-content/uploads/library/govtcomp_report.pdf

Stroud, D. S. (2015). The Bottom of the Iceberg: Unpublished Opinions. *Campbell Law Review, 37*(2). https://scholarship.law.campbell.edu/cgi/viewcontent.cgi?article=1601&context=clr

Stuessy, M. M. (2019). Federal Records: Types and Treatments. *CRS In Focus, 11119*. https://crsreports.congress.gov/product/pdf/IF/IF11119

Summers, E. (2013). *The Web as a Preservation Medium*. Inkdroid. https://inkdroid.org/2013/11/26/the-web-as-a-preservation-medium/

Summers, E. (2020). *Legibility Machines: Archival Appraisal And The Genealogies Of Use*. University of Maryland]. https://raw.githubusercontent.com/edsu/diss/master/diss.pdf.

Sweeney, M. (1992). Guide to United States Government Legislative Documents in the Library of Congress. https://web.archive.org/web/20010417041538/https://www.loc.gov/rr/news/us.legis.docs.html

Tallman, N. (2021). A 21st Century Technical Infrastructure for Digital Preservation. *Information Technology and Libraries*, 40(4). https://doi.org/10.6017/ital.v40i4.13355

Tambouris, E., Manouselis, N., & Costopoulou, C. (2007). Metadata for digital collections of e-government resources. *The Electronic Library*, 25(2), 176–192. https://doi.org/10.1108/02640470710741313

Tenopir, C., Hughes, D., Allard, S., Frame, M., & Birch, B. (2015). Research Data Services in Academic Libraries: Data Intensive Roles for the Future. *Journal of eScience Librarianship*, 4(2). https://doi.org/10.7191/jeslib.2015.1085

Theimer, K. (2008). NARA and the web harvest: a discussion of the issues. *ArchivesNext*. https://web.archive.org/web/20090615030346/http://www.archivesnext.com/?p=137

Thibodeau, K. (2002). Overview of Technological Approaches to Digital Preservation and Challenges in Coming Years. In *The State of Digital Preservation: An International Perspective* (pp. 4-31). Council on Library and Information Resources. https://www.clir.org/pubs/reports/pub107/thibodeau/

Thomas Jefferson Foundation. ([n.d.]). *If we are to guard against ignorance. (Spurious Quotation)*. https://www.monticello.org/site/research-and-collections/if-we-are-guard-against-ignorance-spurious-quotation

Thomas, G. (2022a). [personal e-mail].

Thomas, G. (2022b). [personal e-mail].

Thompson Reuters. (2022). *Federal Reporter, 3d (National Reporter System)*. https://store.legal.thomsonreuters.com/law-products/Reporters/Federal-Reporterreg-3d-National-Reporter-System/p/100000584

Tillett, B. (2004). What is FRBR? http://www.loc.gov/cds/downloads/FRBR.PDF

Toffler, A. (1971). *Future shock*. Bantam Toronto. https://www.worldcat.org/title/14076594

Truesdell, C. B. (2010). Guide to Historical Research with Government Publications, 1789–1989. *Indiana Libraries, 29*(1), 94–112. https://journals.iupui.edu/index.php/IndianaLibraries/article/download/445/412

US Court of Claims. (1970). Jurisdiction of United States Court of Claims. https://catalog.gpo.gov/F/?func=direct&doc_number=000779714&local_base=GPO01PUB

US Court of Claims. (1982). *Cases decided in the United States Court of Claims*. https://catalog.gpo.gov/F/?func=direct&doc_number=000206937&local_base=GPO01PUB

US Bankruptcy Court (Colorado). (1995). Federal rules of bankruptcy procedure and local bankruptcy rules and forms. https://catalog.gpo.gov/F/?func=direct&doc_number=000438688&local_base=GPO01PUB

US Congress. House. Committee on House Administration. (2017). Transforming GPO for the 21st century and beyond : hearing. https://purl.fdlp.gov/GPO/gpo86708

US Court of International Trade. (1984). *Reports*. https://catalog.gpo.gov/F/?func=direct&doc_number=000327141&local_base=GPO01PUB

US Department of Defense. (2022a). *DOD Websites*. https://www.defense.gov/Resources/Military-Departments/DOD-Websites/

US Department of Defense. (2022b). *Military Departments*. https://www.defense.gov/Resources/Military-Departments/

US Department of State. (1995). *Department of State Foreign Affairs Network (DOSFAN).* https://clintonwhitehouse4.archives.gov/WH/Cabinet/html/Department_of_State.html

US Department of State. (2021). *Request for Records Disposition Authority, Bureau of Public Affairs Department of State Public Website, Schedule Number: N1-059-09-004.* https://www.archives.gov/files/records-mgmt/rcs/schedules/departments/department-of-state/rg-0059/n1-059-09-004_sf115.pdf

US Department of State, Office of the Historian. (2025). *The Foreign Relations of the United States (FRUS) series.* https://history.state.gov/historicaldocuments

US Digital Services. (2023). United States Digital Service. https://www.usds.gov/

US Environmental Protection Agency. (2022). EPA Web Archives Home. https://web.archive.org/web/20220402091214/https://archive.epa.gov/

US General Accounting Office. (1990). NASA is not properly safeguarding valuable data from past missions. *GAO/IMTEC-90-1.* https://apps.dtic.mil/sti/tr/pdf/ADA280705.pdf

US General Accounting Office. (2001). Electronic Dissemination of Government Publications. *GAO Report.* http://www.gao.gov/assets/240/231303.pdf

US House of Representatives, Office of the Law Revision Counsel. (2016). *United States Legislative Markup: User Guide for the USLM Schema.* https://github.com/usgpo/uslm/blob/main/USLM-User-Guide.pdf

US Law Library of Congress. (2025). Web Archives. https://www.loc.gov/research-centers/law-library-of-congress/collections/web-archives/

US Library of Congress. (2007). *Saving the World Wide Web.* https://digitalpreservation.gov/series/challenge/web_harvest_challenge.html

US Library of Congress. ([n.d.]). *United States House of Representatives Web Archive.* https://www.loc.gov/item/lcwaN0003449/

US Library of Congress. ([n.d.]). *Web Archives.* https://www.loc.gov/web-archives/

US Library of Congress. (2012). Bibliographic Framework as a Web of Data: Linked Data Model and Supporting Services. https://www.loc.gov/bibframe/pdf/marcld-report-11-21-2012.pdf

US Library of Congress. (2016). Government Publications - United States. *Collections Policy Statements and Overviews.* https://www.loc.gov/acq/devpol/govus.pdf

US Library of Congress. (2017). Web Archiving Supplementary Guidelines. *Collections Policy Statements.* https://www.loc.gov/acq/devpol/webarchive.pdf

US Library of Congress. (2019). Digital Strategy: the FY 2019-2023 Digital Strategic Plan of the Library of Congress. *Version 1.1.2.* https://loc.gov/static/portals/digital-strategy/documents/Library-of-Congress-Digital-Strategy-v1.1.2.pdf

US Library of Congress. (2020a). *WARC, Web ARChive file format.* Sustainability of Digital Formats: Planning for Library of Congress Collections. https://www.loc.gov/preservation/digital/formats/fdd/fdd000236.shtml

US Library of Congress. (2020b). *Law Library of Congress Signs Preservation Steward Agreement with Government Publishing Office.* News from the Library of Congress. https://www.loc.gov/item/prn-20-069/?loclr=ealn

US Library of Congress. (2020c). JATS, Journal Article Tag Suite, NISO Z39.96. https://www.loc.gov/preservation/digital/formats/fdd/fdd000451.shtml

US Library of Congress. (2020d). PDF/A Family, PDF for Long-term Preservation. https://www.loc.gov/preservation/digital/formats/fdd/fdd000318.shtml

US Library of Congress. (2022a). Annual Report of the Librarian of Congress. *for the fiscal year ending Sept. 30, 2021.* https://www.loc.gov/static/portals/about/reports-and-budgets/documents/annual-reports/fy2021.pdf

US Library of Congress. (2022b). *About the Serial and Government Publications Division*. https://web.archive.org/web/20210322073151/https://www.loc.gov/rr/news/brochure.html?loclr=bloglaw

US Library of Congress. (2023). Collections with Web Archives. https://www.loc.gov/web-archives/collections/

US Library of Congress. (2025). Preservation Directorate. https://www.loc.gov/preservation/about/org.html

US Library of Congress. (2025a). Web Archiving: Frequently Asked Questions. https://www.loc.gov/programs/web-archiving/about-this-program/frequently-asked-questions/

US Library of Congress, Federal Research Division. (2018). *Disseminating and Preserving Digital Public Information Products Created by the U.S. Federal Government: A Case Study Report*. https://purl.fdlp.gov/GPO/gpo108102

US National Commission On Libraries And Information Science. (1990). *Principles of Public Information*. https://web.archive.org/web/19970630120453/http://www.nclis.gov/info/pripubin.html

US National Commission on Libraries and Information Science. (1999). *Report on the Assessment of Electronic Government Information Products, commissioned by the United States Government Printing Office Superintendent of Documents*. U.S. Government Printing Office. https://www.govinfo.gov/content/pkg/GOVPUB-GP3-2af80074ebd6803602c2dba265825c8f/pdf/GOVPUB-GP3-2af80074ebd6803602c2dba265825c8f.pdf

US National Commission On Libraries And Information Science. (2000). Public Sector/Private Sector Interaction In Providing Information Services: Report To The NCLIS From The Public Sector/Private Sector Task Force. https://catalog.hathitrust.org/Record/007402468

US National Commission On Libraries And Information Science. (2001). A Compilation Of Recent Federal Statutes Pertaining To Public Information Dissemination. *A Comprehensive Assessment Of Public Information Dissemination, Final Report v.4 [Draft]*. https://web.archive.org/web/20080916052705/http://www.nclis.gov/govt/assess/ComprehensiveAssessment.Volume4.draft.pdf

US National Library of Medicine. (2012). NLM Archiving and Interchange Tag Set. https://dtd.nlm.nih.gov/archiving/index.html

US Office of the Federal Register. (2020). *United States Government Manual*. National Archives and Records Administration. https://www.govinfo.gov/app/details/GOVMAN-2020-11-10

US President. (2023). Economic Report Of The President. https://bidenwhitehouse.archives.gov/wp-content/uploads/2023/03/ERP-2023.pdf

US Senate, Subcommittee of The Committee On Appropriation. (2019). Legislative Branch Appropriations For Fiscal Year 2020. https://www.govinfo.gov/app/details/CHRG-116shrg89104851/CHRG-116shrg89104851

US Supreme Court. (2025). [United States reports] Official reports of the Supreme Court. https://catalog.gpo.gov/F/?func=direct&doc_number=000834059&local_base=GPO01PUB

Ulrich, H., Kock-Schoppenhauer, A.-K., Deppenwiese, N., Gött, R., Kern, J., Lablans, M., Majeed, Stöhr, M. R., Stausberg, J., Varghese, J., Dugas, M., & Ingenerf, J. (2022). Understanding the Nature of Metadata: Systematic Review. *Journal Of Medical Internet Research*, *24*(1). https://doi.org/10.2196/25440

United States Internet Council, & International Technology and Trade Associates. (2000). *State of the Internet 2000*. https://web.archive.org/web/20030112191021/http://www.sdnbd.org/sdi/issues/IT-computer/State%20of%20the%20Internet%202000.htm

University of California Berkeley Library. (2023). *UC Berkeley's libraries*. https://www.lib.berkeley.edu/hours

University of California Curation Center. (2018). *The Datamirror.org Experiment: Preservation Assurance for Federal Research Data*. https://uc3.cdlib.org/2018/07/03/the-datamirror-org-experiment-preservation-assurance-for-federal-research-data/

University of California San Diego Library. (2023). Chronopolis Technologies. https://library.ucsd.edu/chronopolis/technologies/index.html

University of Illinois. (2023). *Library Directory*. https://www.library.illinois.edu/geninfo/library-directory/

University of Texas at Austin. (2023). *Libraries*. https://www.utexas.edu/research/libraries

USAFacts. (2025). *About USAFacts*. https://usafacts.org/about-usafacts/

USAgov. (2025). National Laboratories. https://www.usa.gov/federal-agencies/national-laboratories

Van de Sompel, H., Nelson, M. L., Sanderson, R., Balakireva, L. L., Ainsworth, S., & Shankar, H. (2009). Memento: Time Travel for the Web. *arXiv:0911.1112 [cs]*. http://arxiv.org/abs/0911.1112

Wahid, N., Warraich, N. F., & Tahira, M. (2018). Mapping the cataloguing practices in information environment: a review of linked data challenges. *Information and Learning Science, 119*(9/10), 586–596. https://doi.org/10.1108/ILS-10-2017-0106

Wallace, D. A. (2001). Electronic Records Management Defined by Court Case and Policy. *The Information management journal, 35*(1), 4. https://www.proquest.com/openview/a3382bff53384e46390fe09d6018f6e1/1

Warnock, J. E. (1995). The Camelot Project. https://web.archive.org/web/20020124103219/http://www.planetpdf.com/planetpdf/pdfs/warnock_camelot.pdf

Waters, D., & Garrett, J. (1996). Preserving Digital Information: Report of the Task Force on Archiving of Digital Information commissioned by The Commission on Preservation and Access and The Research Libraries Group. http://www.clir.org/pubs/abstract/pub63.html

Watson, L. C. (2010). *An Evaluation Of PACER*. University of North Carolina at Chapel Hill]. https://cdr.lib.unc.edu/downloads/bc386n994

Weinberger, D., & Searls, D. (2015). *Hear, O Internet*. Cluetrain. https://cluetrain.com/newclues/

Westin, M. (2024). Indexing the information age. *Aeon*. https://aeon.co/essays/the-birth-of-our-system-for-describing-web-content

Wikipedia. (2025). Bulletin board system. https://en.wikipedia.org/wiki/
Bulletin_board_system

Wilkinson, M. D., Dumontier, M., Aalbersberg, I. J. J., Appleton, G.,
Axton, M., Baak, A., Blomberg, N., Boiten, J.-W., da Silva Santos, L.
B., Bourne, P. E., Bouwman, J., Brookes, A. J., Clark, T., Crosas, M.,
Dillo, I., Dumon, O., Edmunds, S., Evelo, C. T., Finkers, R., Mons, B.
(2016). The FAIR Guiding Principles for scientific data management
and stewardship. *Scientific Data, 3*(1), 160018. https://doi.org/
10.1038/sdata.2016.18

Williams, T. (2020). Out of pace with reality? PACER's flaws run counter
to original purpose of increasing access to law. *ABA Journal.* https://
www.abajournal.com/web/article/out-of-pace-with-reality-pacer

Wilson, A. (2007). Significant properties report. *InSPECT work package,
2.* https://significantproperties.kdl.kcl.ac.uk/
wp22_significant_properties.pdf

Woodley, M. S. (2005). *DCMI Glossary.* Using Dublin Core. https://
www.dublincore.org/specifications/dublin-core/usageguide/glossary/

Woods, K. (2012). *The SUDOC Virtualization Project (2007-2010).*
https://web.archive.org/web/20121118142423/svp.soic.indiana.edu/
svp/

Woods, K., & Brown, G. (2009). Creating Virtual CD-ROM Collections.
International Journal of Digital Curation, 4(2), 184–198. https://
doi.org/10.2218/ijdc.v4i2.107

Woods, K. A. (2010). *Preserving Long-Term Access To United States
Government Documents In Legacy Digital Formats.* Indiana
University]. https://web.archive.org/web/20130127122018/http://
www.digpres.com/publications/Kam-Woods-Dissertation-Pre.pdf

World Wide Web Consortium. (2008). *Web Content Accessibility
Guidelines (WCAG) 2.0.* https://www.w3.org/TR/WCAG20/

Wyatt, P. (2024). Files inside PDF. https://pdfa.org/files-inside-pdf/#

Young, S. (2006). Guide to CRS Reports on the Web. *LLRX.* https://
web.archive.org/web/20100805044410/https://www.llrx.com/
features/crsreports.htm

Zegart, A. B. (2022). American Spy Agencies Are Struggling in the Age of Data. *Wired*. https://www.wired.com/story/spies-algorithms-artificial-intelligence-cybersecurity-data/

Zhu, X. (2012). *Access to Digital Case Law in the United States: A Historical Perspective*. In *IFLA 2012*. https://www.ifla.org/past-wlic/2012/193-zhu-en.pdf

Zierau, E. (2017). OAIS and Distributed Digital Preservation in Practice. *Proceedings of the 14th International Conference on Preservation of Digital Objects*. http://www-archive.cseas.kyoto-u.ac.jp/ipres2017.jp/wp-content/uploads/14Eld-Zierau.pdf

Cover

Cover Design by Deborah Yun Caldwell

Cover Image: Architect of the Capitol. Cropped detail of the Frieze of American History surrounding the Apotheosis of Washington in the United States Capitol Rotunda. This image is in the public domain and does not imply endorsement by the Architect of the Capitol or the United States Congress.

About the Authors

James A. Jacobs ("Jim") is Librarian Emeritus, University of California San Diego where he served as Data Services Librarian from 1985 to 2006. He co-taught the ICPSR workshop, "Providing Social Science Data Services: Strategies for Design and Operation" from 1990 to 2012. He served as a technical consultant to the Center for Research Libraries in their audits and certifications of the Trusted Digital Repositories of HathiTrust, Portico, and LOCKSS, among others. He is a co-founder of Free Government Information (freegovinfo.info).

James R. Jacobs is the US Government Information Librarian at Stanford University Libraries where he supports the research needs of the university, and works on both traditional collection development as well as digital projects like LOCKSS-USDOCS and Web harvesting, including the End of Term Archive and FOIA. He received his MSLIS in 2002 from the University of Illinois at Urbana-Champaign. His passion and expertise lie in the realm of digital preservation, digital collection development, FOIA, and the expansion of the public domain and information commons. He is a co-founder of Free Government Information (freegovinfo.info) and Radical Reference (radicalreference.info).